The Science of Structures and Materials

THE SCIENCE OF STRUCTURES AND MATERIALS

J. E. Gordon

**SCIENTIFIC
AMERICAN
LIBRARY**

A division of HPHLP
New York

Library of Congress Cataloging-in-Publication Data

Gordon, J. E. (James Edward), 1913–
    The science of structures and materials/J. E. Gordon.
      p.   cm.
    Bibliography: p.
    Includes index.
    ISBN 0-7167-5022-8
    1. Strength of materials.   2. Structures, Theory of.   I. Title.
TA405.G62 1988   87-35412
620.1'1—dc19   CIP

Printed in the United States of America.

Scientific American Library
A Division of HPHLP
New York

Distributed by W. H. Freeman and Company.
41 Madison Avenue, New York, New York 10010 and
20 Beaumont Street, Oxford OXI 2NQ, England

  2 3 4 5 6 7 8 9 0 KP 6 5 4 3 2 1 0 8 9 8

This book is number 23 of a series.

Contents

To my grandchildren
Katerine, Nicholas, and Francis
    who are likely to live to
    experience the consequences
    of this kind of thing

Preface

Even a moment's thought must show that the possession of adequate mechanical properties—strength and toughness, for example, stiffness, or flexibility—is essential for the existence and continuity of all forms of life. This is true not only for plants and animals and human beings but for nearly all the technological works of man from the stone age to the space age.

Yet the whole subject of the mechanical properties of materials and structures, living or artificial, has gone largely ignored, both by ordinary people and by the great majority of scientists. It is usually passed over very briefly in school science courses, and it is apt to be brushed aside, as trivial or irrelevant, in most university courses in biology and medicine, physics and chemistry, not to mention history. Naturally, engineers have had to study these matters, but until very recently, engineers have seldom talked to doctors or biologists, let alone to historians.

Quite lately, perhaps because of "lateral thinking," the whole picture has been changing. The engineers, the doctors, and the biologists are at last beginning to get together and to explore the new and rapidly growing science of biomechanics, that is, the study of the mechanical behavior of living materials and structures. As a result, new and perhaps revolutionary technological materials, based on biological models, are now being developed.

No doubt this new approach will benefit medicine, especially orthopedics, in various ways. But its more revolutionary effects are likely to be felt in technology—not so much, perhaps, in "high technology" as in the technologies of everyday life: in buildings, in vehicles, in machinery. A revolution is arising where it is least expected: in established and familiar things, and just for this reason it will probably have a radical effect both upon our lives and upon our ways of thinking. The revolution deserves the attention of not only scientists, engineers, and ordinary people but of industrialists, economists, historians, and politicians as well.

For their help in writing this book, I must thank many of my colleagues, especially Dr. Giorgio Jeronimidis and Dr. Julian Vincent. My secretary, Mrs. Jean Collins, has been a great help, and my wife has displayed admirable fortitude during the writing and production of this book.

J. E. GORDON
UNIVERSITY OF READING, ENGLAND
DECEMBER 1987

The Science of Structures and Materials

1

THE STUDY OF STRENGTH

A Cinderella science

In all natural things there is something of the marvelous.
There is a story which tells how some visitors once wished
to meet Heraclitus, and when they entered and saw him
in the kitchen, warming himself at the stove, they hesitated.
But Heraclitus said, "Come in; don't be afraid.
There are gods even here."

ARISTOTLE
De partibus animalium

The study of the strength of structures and materials, which is now fashionable, has been a Cinderella among the serious sciences. Like medicine in an earlier age, for many years the study of strength tended to be a pragmatic endeavor much beset by superstitions and half-truths. Like medicine in any age, the science of strength is apt to be complicated and difficult: simplistic theories can often be dangerous. The strength of structures and materials is a subject that affects us all closely in many different ways. Perhaps for this reason many people would rather not know too much about it. The solar system, for instance might seem an interesting, romantic, and safely remote topic, whereas why bones and airplanes break are questions that make us feel uncomfortable.

For many years a knowledge of materials and structures was to a large extent the province of craftsmen—of shipwrights, architects, and "practical" engineers—and this knowledge was mostly traditional in character. Cooperation with academic scientists, when it did occur, was not always fruitful. Attempts by theoreticians to predict the actual strengths of real material from the known strengths of their chemical bonds did not command much respect among engineers, for these estimates were frequently in error by a factor of a hundred or more. Mathematicians did cut some teeth on the design of large iron and steel structures early in the nineteenth

Stave church at Gol, Norway, built entirely of wood. Artisans working with traditional lore can construct elegant and highly sophisticated buildings without a close familiarity with technical training.

century, but there is a long concomitant history of disasters in which ships and bridges broke despite all calculations, sometimes with heavy loss of life.

After some of these initial experiences with theory, engineers were understandably inclined to return to their handbooks and Codes of Practice. The academics retaliated by intimating that the whole question of how to support a load was intellectually trivial and of little scientific interest anyway. This was no pure science. In fact, many universities strenuously resisted the establishment of departments of engineering until surprisingly recently.

For many years, and well into this century, respectable academic tradition favored "pure" specialization—the more specialized, the better. The hope or the fear of each practitioner was that, like nuclear physics, his field of study would turn out to be of practical importance. The flow of ideas thus ran from the pure to the applied—as illustrated in the following anecdote. The British statesman W. E. Gladstone once paid a visit to Michael Faraday's laboratory in London. The scientist showed the cabinet minister some of his experiments with electromagnetism and electrostatic induction. Gladstone, though a sound classical scholar, was not endowed with much technical imagination: "Very interesting indeed," he remarked, "but tell me, Mr. Faraday, of what *use* is this electricity?" "Sir," said Faraday, "you will soon be able to tax it!" (This story is also told of Sir Robert Peel, but I find the Gladstone version more probable.)

The study of structures, however, often works the other way around: the flow of ideas is as often from practice to theory as from theory to practice. In some respects this field is the opposite of a pure science, since it is essentially purposive. The notion of a "pure" structure—a structure with no application—is not a very fruitful one. Also, the field today is hardly a specialty. Both living and artificial structures exist in enormous variety; their study is essentially catholic and comparative in its nature.

Nowadays, progress is being made by bringing together experts in widely different fields: engineers and shipbuilders; botanists, zoologists, and doctors; mathematicians and physicists; chemists, metallurgists, and crystallographers. Until quite recently many of these specialists resisted, often very strongly, attempts to get them to communicate with one another. Among the few exceptions was Sir John Charnley (1911–1983), the leading orthopedic surgeon in England in his day, and the holder of a degree in engineering as well as his medical degrees. Most biologists showed little interest in quantitative analysis of the structural efficiency of living things; in fact, structural efficiency itself was regarded as incidental, rather than as critically important in the Darwinian struggle. Engineers tended to despise natural material; it was self-evident to them that metals like steel and aluminum were better. Furthermore, specialization was so complete that metallurgists, immersed in their eutectics (studies of melting point) and phase diagrams, did not really want to know about the engineering contrivances made possible by their alloys. Both in technology and in academic research, the distinction between *materials* and *structures* tended to be rigid.

Although there are still some intellectual hermits, communications have improved a great deal during the last few years. This has been due partly to the rise of the new science of biomechanics, the study of the mechanical properties of living organisms. Another reason is the impact of new materials—such as plastics and composites—on industry and on technological thinking. Then again, old materials such as wood are making a comeback in engineering.

Perhaps, too, we are becoming more liberal in our ideas and more prepared to accept that science does not have to be grand or remote, to dwell up among the stars or way down among the atoms. Like the gods and Cinderella, it may live in the kitchen. Although it is unashamedly useful, the science of the strength of structures and materials throws much philosophical light upon the evolution of life, upon the progress of technology, and upon history in general—nor is it lacking in intellectual depth and rigor.

So this book is concerned with several different disciplines and with more than one culture. And it is about *both* materials and structures.

WHAT IS A STRUCTURE?

A *structure* is often defined as "any assemblage of materials which is intended to sustain loads." The key word here is "intended." Many things that could scarcely be described as purposive structures in terms of human design or biological evolution can resist mechanical forces in a fortuitous way: mountains, planets, and stars are generally able to withstand the gravitational effects of their own mass. Although our umbrella definition of structure is not quite large enough to cover these natural entities, they have many similarities with purposive structures. In fact, the analysis of such phenomena as landslips, earthquakes, glaciers, or black holes in space is generally a spinoff from the science of structures.

The structures in which we are interested here are those that exist to serve the ends of life in some way. Only a minority of them are made by human beings; the rest are products of biological design. Before there was life in the world there was no such thing as a purposive structure—only heaps of sand and rock. The land surfaces of this planet must have looked much like the surface of the moon does today, while the seas were empty wastes of water. Almost the first thing Nature had to do when she invented life was to devise structures to contain it. (I capitalize Nature and write teleologically here to spare myself a good deal of circumlocution; philosophers and theologians may make of this what they choose.) Even the most primitive unicellular organisms had to be enclosed and protected by cell membranes that were both flexible and strong and yet capable of accommodating cell division during reproduction. As evolution advanced and life became more competitive, the structural requirements became more sophisticated. The majority of living tissues have to carry mechanical loads of one kind or another; some tissues, like muscles, have to *apply* loads as well, and to change shape while doing so. Most plants and trees are designed to grow tall and stand up to being buffeted by the wind. Animals often need to be able to swim or run fast, and many of them must have strong bones, tendons, teeth, or claws. Humbler things like snails survive by virtue of their strong shells.

These mechanical properties are *primary*—not secondary—characteristics of biological tissues, even though in the economy of Nature structures like plant stems and animal skins have to perform other functions besides those associated with strength. Biology places a great premium on strength and mechanical safety, while requiring also that tissues be light in weight and metabolically efficient. In fact many trees and birds and animals are very efficient structures indeed. Most of the works created by humans must resist mechanical forces of various kinds, so they qualify as structures, too.

History has depended, more than most historians have acknowledged, upon the development of better materials and structures. The Roman Empire was richer and better organized in most things than medieval Europe, and the ancients built bigger ships. But, like St. Paul's vessel (Acts of the Apostles, chapter 27), they leaked catastrophically and could not keep off a lee shore in a gale because of their structural inadequacies. Consequently ancient seafarers were more or less confined to fair-weather voyages during the Mediterranean summer, and they dared not venture into the great oceans of the world. The discovery and colonization of America had to await the evolution of ships with strong rigs and watertight hulls.

For another example, apparently the Chinese invented gunpowder several centuries before it was known in the West, but explosives had limited military value in the absence of safe and effective gun barrels. The art of casting cannon, developed in the West during the fourteenth century as an offshoot of the technology of casting church bells, allowed the full lethal potency of gunpowder to be realized.

In a later age the advent of railroads and automobiles depended upon supplies of cheap and reliable iron and steel. Fortunately, in cars and trains there is a good deal of latitude for inefficiency. Weight is not absolutely

A structure is any assemblage of materials intended to sustain loads. Plants such as tree are structures, designed to grow tall and stand up to strong winds.

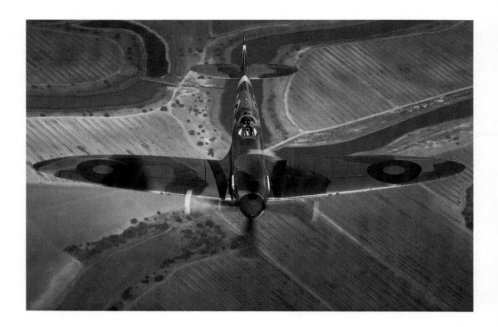

An RAF Spitfire single-seat fighter.

critical, and mistakes, ignorance, or laziness in design and manufacture can often be covered by making the parts unnecessarily thick and heavy. The development of aircraft and rockets, on the other hand, has forced us to pay more serious attention to the science of materials and structures. In aerospace weight is a luxury and breakage is likely to mean death. As we pursue newer technologies, therefore, we are "getting back to Nature."

Because our subject and its applications are so broad, this new knowledge is potentially revolutionary. Until lately military specialists, politicians, economists, businesspeople, and many others have been ready to come to terms with innovations that have appeared, like fairy godmothers, from outside traditional technologies, such as the electric telegraph in the past or computers today. Despite their obvious importance, they might still be rudely described as gadgets. People are more reluctant to envisage radical changes in the basic and customary things of life, things that most of us do not often think about although they represent very considerable industries.

In the writer's opinion the modern approach to materials and structures is not likely to result in a plethora of exciting new gadgets. Its impact will probably be less dramatic in the area of "high" technology than in the traditional trades and businesses, where it may prove to be a major time bomb. For instance, we have become accustomed to regarding the steel industry as residing more or less permanently at the heart of the economy. The manufacture of steel has indeed had a massive influence on the development of technology during the last century or so. Steel has been not just a material, but also a social and historical phenomenon. In many established industries the division of labor and other business arrangements whose patterns have been sanctified by Adam Smith and Henry Ford derive in large measure from the physical characteristics of metals like steel.

However, the steel industry may soon go the way of the steam engine, another child of the Industrial Revolution. Materials that will replace steel in the future are unlikely to be direct substitutes, technically, economically, or socially. A number of important apple carts will probably be upset. Students of prehistory refer to the Stone Age and the Bronze Age: sooner than we think historians may be writing about the Steel Age.

Building construction has long been a rather conservative activity, which makes it particularly vulnerable to change. Houses and other buildings are getting more and more expensive to construct, so we may soon see cheaper and less conventional ways of making them. These innovations are likely to involve disconcerting changes in the deployment of labor and capital. Furthermore, the appearance of buildings will probably change drastically, too, in which case our Main Streets and our suburbs may give some of us a bit of a shock.

Although the subject of strength affects everybody, its study has been made unreasonably difficult for the general reader because so much of the literature contains page after page of mathematics. The introduction of some numbers is unavoidable, but we shall get by here with only the simplest kind of algebra.

2

THE LANGUAGE OF STRENGTH

The Muddle about Stress and Strain

The Centipede was happy quite,
Until the Toad in fun
Said "Pray, which leg goes after which?"
And worked her mind to such a pitch,
She lay distracted in the ditch
Considering how to run.

MRS. EDMUND CRASTER (d. 1874)

In many sciences progress has periodically awaited the introduction of clever mathematics or sophisticated instruments, such as telescopes or electron microscopes. The stumbling blocks for physicists and astronomers are generally not conceptual ones; these scientists tend to know what their actual problems are even if they do not have the means to solve them. Furthermore, although the professional language of physicists and astronomers might seem obscure to the layperson, it is relatively objective and free of complications introduced by the subconscious. Studies of stars and atoms may have progressed as well as they have because such entities are so remote from our everyday experience that we are not burdened by too many preconceptions about them.

Technological innovations and mathematical leaps forward have played their part in materials science and in the study of structures, but the real difficulties and the real advances have been conceptual ones. The difficulties have certainly been due in part to the long working familiarity of humans with structures and materials in everyday life. Both people and animals have grappled with problems in the mechanics of solids since before the beginnings of rational thought. The ''higher'' animals are saturated with deeply held subjective ideas concerning the physical elements of their environment. Throughout evolution these feelings have generally been protective and beneficial. For example, animals rarely break the things on which they climb. Birds perch on very thin twigs, sometimes on ones that are clearly rotten—yet branches never seem to give way beneath the weight of birds.

We human beings may not be as clever as birds at estimating the strength of objects in our environment, but most of us are quite good at it. We do not skate on thin ice, lean against collapsible structures, venture across rickety bridges, or drive into mudholes. No doubt our avoidance of these errors is partly the result of species heredity and partly learned from painful experience in childhood. In any case, our decisions tend to be subconscious. We may be in the emotional situation of the tightrope walker who is reluctant (like the centipede in Mrs. Craster's verse) to analyze her performance scientifically lest she suddenly lose confidence and hurt herself.

This subjective, intuitive approach to estimating strength is reflected in the nature of our language. In ordinary speech the meanings of words like ''strong,'' ''tough,'' ''hard,'' and ''rigid'' are usually vague, and they are often used interchangeably. Indeed one has only to consult a dictionary or talk to a construction worker to appreciate the extraordinary confusion of words and ideas associated with the concept of strength. Nor is this confusion peculiar to the English language; experience of teaching engineering to foreign students has shown me that other modern languages are, if possible, even more muddled and ambiguous.

Feelings about the physical elements of the environment may be both protective and beneficial. Birds seldom break the twigs on which they land.

ANCIENT AND MEDIEVAL CONSTRUCTION

Ancient Greek and Latin were just as careless with the everyday terms associated with structures and materials as modern languages are. *Tensio* in Latin, for example, can mean either tension or extension: "Ut tensio, sic vis" ("As the extension, so the force"). The linguistic imprecision of the ancients in this regard makes their great skill at measurement and symmetrical construction all the more remarkable. They erected an aesthetically magnificent array of large buildings and other artifacts—but these things were (with the exception of certain weapons of war) structurally inefficient. Apparently both the intellectual Greeks and the practical Romans suffered from a social snobbery about technology which prevented them from making a serious attempt to break through the linguistic and conceptual barriers.

For instance, the Parthenon in Athens, built around 440 B.C., achieves a geometrical precision that has almost certainly never been approached by any modern building. A fantastic amount of thought, care, and instrumentation must have gone to ensure dimensional accuracy. However symmet-

The Parthenon at Athens.

The Treasury of the Athenians at Delphi, built around 490 B.C. Note the cracked architrave.

rical and beautiful the result, though, the structure is not sound. Many of the architraves have cracked, being clearly inadequate as beams. That the Greeks were somewhat backward when it came to engineering is not surprising in light of the fact that even Archimedes (circa 287–212 B.C.), the greatest Greek engineer, was ashamed of his profession. He refused to write a handbook on engineering because, according to Plutarch, "he looked upon the work of an engineer and everything that ministers to the needs of life as ignoble and vulgar." Heraclitus (fl. 510 B.C.), with his ideas about gods living in the kitchen, was exceptional or perhaps old-fashioned. Aristotle (386–322 B.C.) quotes him (see the Epigraph to Chapter 1) only to justify the study of small and unattractive animals.

The Romans did not take quite this point of view but, in spite of their considerable technical achievements, their approach to design problems was not analytical in any modern sense. Vitruvius's textbook *De architectura* (circa 30 B.C.) is deservedly famous, but the treatment is descriptive and pragmatic. Other Roman writers on the subject are no more rigorous. The ancients must have relied much more than they cared to admit upon the skills and experience of the illiterate craftsmen whom they despised.

On land, structural accidents seem to have been fairly rare except in the case of the *insulae* or tall apartment blocks that sprang up in Rome at the time of Augustus (63 B.C.–A.D. 14). These fell down so often that the emperor had to pass a law restricting their height to 66 feet (20 meters). Otherwise, the Romans apparently built their bridges, forts, and aqueducts with what we would now call very large margins of safety.

At sea engineering mishaps were far more common, as mentioned in Chapter 1. Classical literature is full of accounts of shipwrecks, and the ancients understandably regarded the sea and sea voyages with horror. But discounting errors of seamanship and pilotage, most shipwrecks were due to mechanical weaknesses in hulls or rigs. The construction of ancient ships was, in fact, so appalling that it seems strange that the ancient philosophers paid little attention to the problems of ship construction.

During the early middle ages in western Europe, formal learning decayed, but structural craftsmanship made conspicuous advances, especially in shipbuilding. The ships of the Norsemen were certainly stronger, more watertight, and more seaworthy than Roman ships. Consequently the Norsemen were able not only to discover Iceland, Greenland, and parts of North America, but also to maintain some sort of communication with their colonies there across one of the stormiest oceans in the world.

By the fifteenth century some of this structural virtue seems to have infected Spanish, Dutch, and English shipbuilders, with results that are evident in modern maps and history books. However, the wisdom of the shipwrights still resided mostly in the skills of craftsmen, and their modus operandi was seldom quantitative or analytical. For example, the stressing

Trireme under construction by Trieres Project, a joint Anglo-Greek venture. Trieres is the Greek for the Latin triremis (equivalent to three oars). The completed ship (see page 157) is powered by 170 oarsmen.

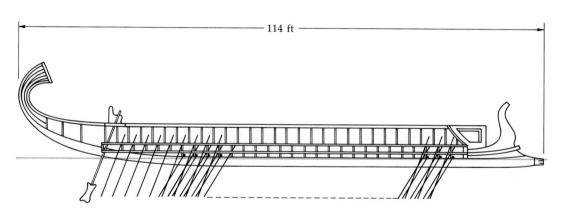

114 ft

Cross sections of a trireme (top) and a Viking ship (bottom).

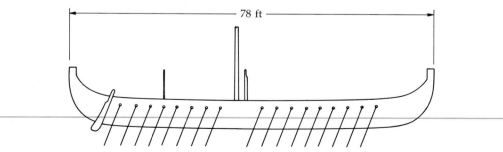

78 ft

of ships' hulls in a scientific manner was not fully accepted practice until around 1900.

On shore, throughout western Europe castles and other fortifications continued to be built very much in the foursquare Roman tradition to maintain the largest possible margin of safety. There are few instances of medieval castles falling down in times of peace, whatever may have happened to them when they were besieged. Gothic churches and cathedrals represent an entirely different architectural philosophy, and one removed as well from ancient traditions. Aesthetically they are the opposite of classical temples. The earth-bound temples convey a feeling of solidity, precision, and permanence. Gothic architecture aspires heavenward; it attempts to convey an ethereal message, using a heavy, brittle material—stone. Extensive analyses of medieval cathedrals have shown that although the master masons built their arched and vaulted structures very near the limits of failure, they certainly did not think or design in any modern way. For many years virtually all construction work, of cathedrals as well as of other buildings and ships, was a monopoly of the craftsmens' guilds. These bodies were highly conservative and no doubt scornful, as craftsmen generally are, of any rational analysis of their work.

A number of the more ambitious Gothic churches and cathedrals did collapse, either during construction or later. The cathedral of Beauvais (begun in the thirteenth century) is a notable example: the tower fell once and the roof fell twice.

Left, interior of the apse of Beauvais Cathedral, France. The building was begun around 1225 and was planned to be the highest Gothic vault (45 meters). The roof collapsed in 1284 and was rebuilt with reinforcements. Only the apse and the transept were completed. Right, polarized light view of stresses in the walls of Beauvais Cathedral. A plastic model of a cross section of the building is loaded with weights to stimulate wind loading. The model is heated to 150°C and allowed to cool. The resulting deformations are then observed in a polariscope. Lowest stresses are in black regions, highest stresses where fringes are closely spaced.

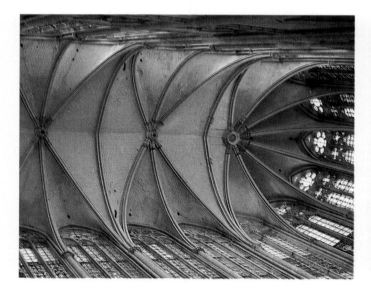

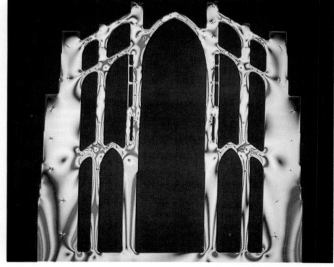

THE RENAISSANCE AND THE BEGINNINGS OF STRUCTURAL SCIENCE

The first recorded evidence of a scientific approach to the problems of strength is found in the notebooks of Leonardo da Vinci (1492–1519). He loaded wires with baskets of sand to test the wires' tensile strength—that is, the size of the smallest load that could break them (*tensile* refers to the pulling action of the load). Interestingly, he seems to have been more concerned with assessing changes in strength due to variations in the *length* of the wire than changes due to variations in its thickness. Although it may seem obvious to us that the static strength of a uniform rope or wire is independent of its length, under sudden or *dynamic* loading. A long rope will stretch farther and so absorb more energy than a short one: it all depends on what is meant by "strength." So Leonardo was not being silly.

Leonardo also attempted, with rather limited success, to analyze the strength of beams, trusses, and columns. His notebooks contain the perceptive—or ominous—remark: "Mechanics is the paradise of mathematical science because here we come to the fruits of mathematics." Although most of this work of Leonardo's remained buried in his notes until modern times, the mathematician Girolamo Cardano (1501–1576) may have had access to these papers, or at least talked on the subject to Leonardo. In any case some of Leonardo's ideas were incorporated into Cardano's books, which are thought to have been read by Galileo.

If anybody deserves the title "father" of the modern science of strength it is the astronomer and physicist Galileo Galilei (1564–1642). He made his contributions somewhat by default. Upon presenting his telescopic observations supporting the Copernican or heliocentric version of the solar system in his famous book *Dialogo dei due massimi sistemi del mondo* (*Dialogue on the Two Major Systems of the World*, 1632), he was summoned to Rome by the Inquisition, which considered his work at variance with Holy Writ, and forced to read his recantation. After his return to Florence late in 1633 he spent the remaining eight years of his life under house arrest at his villa at Arcetri. Banned from astronomy, he turned to the study of materials and structures, which the church considered an innocuous topic.

The subject of strength was not by any means a new field for Galileo. Before astronomy had overtaken his other interests he had written, in 1594, a treatise on mechanics (*Della scienza mechanica*). This work was not published at the time, although it was widely circulated in manuscript. He had apparently been consulted around this date about some problems in shipbuilding, and this had aroused his interest in the behavior of materials.

Now in his seventies, Galileo used his enforced leisure at Arcetri to revive and extend his former interest in the mechanics of strength. Provided he kept off "dangerous" subjects he was free to correspond with

Bending stresses in a beam from Galileo's Discorci e dimonstrazioni matematiche intorno a due nuove scienze, *Leiden, 1638.*

learned men in various parts of Europe, and an extensive collection of his letters has survived. In particular, he exchanged ideas with a French Jesuit priest, Marin Mersenne (1588–1648), who was interested in the strength of the metal wires used in musical instruments.

The result of Galileo's experiments and reflections at Arcetri was the publication in 1638, of *Dialoghi delle due nuove scienze* (*Dialogues Concerning Two New Sciences*). Although this was an extremely influential book, not all its content is scientifically correct. Its impact was due partly to the circumstance that is was the first published work to deal seriously with the subject of the mechanics of strength—although public curiosity about this famous victim of the Inquisition helped to garner the book a wide audience.

In *Dialoghi delle due nuove scienze* Galileo discusses the static strength of rods similar materials but with different cross-sectional areas. He had measured their strength by subjecting them to loading in axial tension—that is, to a straight pull along the axis of rods. He finds that the breaking load is proportional to the area of the cross section; which is what we should expect. He also analyzes the strength of various kinds of beams. Some of his results are correct, but some, such as the strength of the cantilever in the margin, are not. Living and working two generations before Newton, Galileo did not have the benefit of Newton's demonstration that action and reaction are equal and opposite, so that all the forces acting upon and within a body must cancel themselves out.

Galileo's book started a fashion for looking rationally at the problems of mechanical strength, and not only among academics and philosophers. The trend spread to the British Admiralty, to the dismay of fraudulent contractors. The diarist Samuel Pepys (1633–1703) was, in effect, the founder of the British civil service in its modern, comparatively honest form. He could claim to be the first civil servant to make use of the techniques of materials science in the interests of administrative probity. Pepys's diary for 4 June 1662, reads:

> Povey and Sir W. Batten and I by water to Woolwich; and there saw an experiment made of Sir R. Ford's yarn (about which we have lately made so much stir; and I have much concerned myself of our rope-maker, Mr Hughes who represented it so bad) and we found it to be very bad, and broke sooner than, upon a fair triall, five threads of that against four of Riga yarn; also some of it had old stuffe that had been tarred, covered over with new hempe, which is such a cheat as has not been heard of.

In contrast, Pepys's great contemporary, Sir Isaac Newton (1642–1727) did not believe that gods resided in kitchens, let alone in dockyards, and he seems to have shared the Greek contempt for the applied sciences. His chief interests were astronomy and cosmology, mathematics, and what

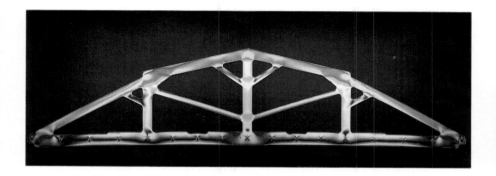

The Sheldonian Theatre, Oxford, designed by Sir Christopher Wren and completed in 1662. Though the building departs from Oxford's Gothic traditions, the roof is constructed very much in medieval fashion, as the polarized light view of the interior of the roof at left shows. The pattern of light shows that the stress was not evenly distributed. At right, exterior of the Sheldonian.

may seem to us rather obscure problems in theology. Nevertheless, Newton's system of mechanics, which was published in his *Principia* in 1687, was destined to play at least as important a part in the design of structures as it did in the analysis of the solar system. Furthermore, Newton's mathematical discoveries, notably the differential calculus, were later to prove invaluable to engineers.

Newton's third law is his statement that action and reaction are equal and opposite, and that all the forces acting within a system must balance out. If a weight presses down on the floor, the floor must press up on the weight with an equal and opposite force. Or, if a weight of 100 pounds is suspended from a rope, then the rope itself has to produce an equal and opposite force of 100 pounds in order to sustain the weight. If the load is increased to 200 pounds, then the rope will have to exert a force of 200 pounds; if the rope cannot produce the necessary force, it will break and the weight will fall. This principle applies throughout every structure, however simple or complicated.

Newton also pointed out that in the case of a stationary weight the load it applies is generated by the action of the Earth's gravitational field upon the mass of the weight. In the case of one or more moving bodies in collision—that is, a bullet and its target, the wind and an obstacle, or two ships ramming one another—the applied force, as Newton also said, is generated by the acceleration or deceleration of the moving masses. These principles are valuable if we want to calculate the forces to which a structure is likely to be subjected, but they say too little about how the necessary forces of reaction are produced in the material of the structure. If we hang a weight on a rope, how and why does the rope resist?

Probably because such problems seldom arose in astronomy, Newton did not show much interest in the matter. The credit for answering this rather important question goes to Newton's bête noire, the British physicist Robert Hooke (1635–1703).

HOOKE: THE ELASTICITY OF SOLIDS

Although intellectually he was nearly as able as Newton, Hooke's views and interests were diametrically opposed to Newton's, and the two men became lifelong enemies. Unlike Newton, Hooke was intensely interested in what went on in kitchens, dockyards, and buildings—the mundane mechanical arenas of life; he was interested, in fact, in almost any practical problem, down to and including the anatomy of fleas.

Nor did Hooke despise craftsmen, and he probably got the inspiration for at least some of his ideas from his friend the great London clockmaker Thomas Tompion (1639–1713), with whom he seems to have had endless discussions about things like springs and pendulums. Naturally, in the seventeenth century Hooke did not have the advantage of modern ideas about the forces generated by deforming interatomic bonds; nevertheless, he may have guessed that something of the kind was happening within the fine-scale structure of a material.

He approached the study of the effects of forces on different materials macroscopically and pragmatically. He took a considerable variety of wires, springs, and wooden beams and loaded them progressively by adding weights to scale-pans. He measured the resulting deflections as well as he could using a pair of compasses. When he plotted the variation of load against the deflection he found that in each case the graph was a straight line. Furthermore, when the load was progressively removed the recovery was also linear, and within the accuracy of his measurements (which was obviously not great), the specimens returned to their original lengths when they were unloaded.

To protect his claims to priority, and perhaps to irritate Newton, Hooke published in 1676 "A decimate of the centesme of the inventions I intend to publish." This prospectus included the heading "The true theory of elasticity or springiness," and this was followed by the anagram "ceiiinosssttuu." Hooke revealed the solution to this anagram in 1679 in his *De potentia restitutiva, or of a spring:* the true theory of elasticity was "ut tensio, sic vis" (which, as we have already seen, means "as the extension, so the force"; in Latin, *tensio* generally means extension or stretching rather than tension in the sense of a force—though elsewhere the Romans, predictably, seem to have muddled the two ideas). Hooke wrote:

> It is very evident that the Rule or Law of Nature in every springing body is, that the force or power thereof to restore itself to its natural position is always proportionate to the distance or space it is removed therefrom, whether it be by rarefaction, or separation of the parts from one another, or by Condensation, or crowding of those parts together. Nor is it observa-

ble in those bodies only, but in all other springy bodies whatsoever, whether metal, wood, stones, baked earth, hair, horns, silk, bones, sinews, glass and the like.

Hooke was saying that a solid can resist an external force only by changing its shape: by stretching if it is subject to a tensile force, or by contracting if it is compressed. His discovery was the logical consequence of Newton's third law. There is, according to Hooke, normally no such thing as an absolutely rigid material or structure. It is true that the deflections under load in masonry and reinforced concrete are generally so small that sophisticated instruments are needed to measure them, but the movement is real enough. On the other hand, trees are visibly bowed by the wind and we have all watched aircraft wing tips moving up and down in flight as we peered at them from the window seat. Springs and rubber bands are only more obvious examples of a behavior that is almost universal.

With an ordinary engineering material such as a metal, the change in length both of the solid as a whole and of the interatomic spacing typically lies between 0.1 percent and 1.0 percent of the original length. That the extension or contraction of the interatomic bonds in a crystalline solid, such as steel, does correspond proportionately to the gross extension or contraction of the material was checked with great care by Norton and Loring in 1941, using X-ray diffraction methods.

Although engineering science did not register the full effects of Hooke's ideas until after the general acceptance of the concepts of stress and strain around 1830, his work has proved in the long run to be of very great importance. It is therefore worth summarizing his three main points and commenting on them in the light of modern knowledge:

1. *A solid material can resist an applied force only by yielding to it—that is, by contracting under a compressive load or by stretching under a tensile one.*
    Hooke explained that this is because the "fine parts" (i.e., the atoms and molecules) of the material are themselves brought closer together ("condensed") or stretched farther apart ("rarefied").

    Broadly speaking, all ordinary solids, whether crystalline or amorphous, follow this rule. Some "active" materials, such as animal muscles and piezoelectric crystals, can produce mechanical forces by other means but energy has to be fed into them. A more important exception occurs with some very soft solids, which like liquids have a surface tension that can resist applied forces to a slight extent. Such materials can play a significant role in biological structures. However, among the ordinary structures and materials of technology, this premise of Hooke's is valid and very important.

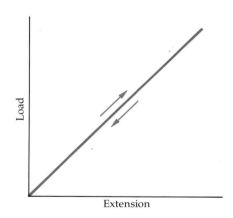

Fully elastic (i.e., recoverable) or Hookean behavior. The degree of extension or compression increases linearly with the size of the load.

2. *Solid materials are elastic. That is, they will recover their original shape and dimensions completely when a load that has been applied to them is removed.*

For inorganic materials such as ceramics and hard metals this is very nearly true, especially with small or moderate loads—for instance, the loads on the springs of cars or watches. Botanical materials sometimes recover only slowly which is why you can see footprints on the grass of a lawn.

Many ductile (malleable) engineering metals are elastic only up to a certain point (see lefthand illustration below). Beyond that point they yield in an inelastic, irreversible way. This is why one can put a permanent bend into a can lid or a piece of wire, and why the automobile manufacturer can press initially flat steel sheets into the curved panels for car bodies.

3. *Hooke's law: In materials and structures the deflection is always proportional to the applied load.*

Hooke's law is strictly true only for ceramics, glass, most minerals, and very hard metals. Ductile metals such as mild steel obey the law up to moderate loads but depart from it irreversibly at large loads.

Most animal tissues do not follow Hooke's law (see righthand figure below). Although their extension under load is fully elastic and reversible, it is by no means linear. Many of our body tissues have a load-extension curve that is approximately J-shaped, as one can feel by pulling on one's lip or on the lobe of one's ear. A gentle tug produces considerable extension, whereas a stronger tug results in relatively little additional extension. As we shall see, this behavior imparts a highly beneficial form of toughness to the animal.

Left, an elastic or plastic behavior shown by ductile metals. At small loads the extension or compression increases linearly with the size of the load, but once a certain load is reached, the metal yields—undergoes a large additional extension or compression—in a way that is mostly irreversible when the load is removed. Right, fully elastic J-curve behavior shown by many animal tissues. The greater the load, the smaller the additional extension or compression. This property confers a desirable toughness to the tissues.

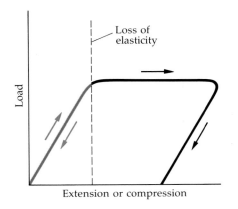

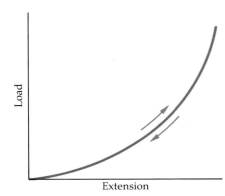

In contrast, the load-extension curve of soft material, such a rubber, which is elastic and also non-Hookean, have, not a J-shaped load-extension curve, but an S-shaped or sigmoid one. This property can make the material dangerously brittle, as one can see by sticking a pin into a blown-up toy balloon. In a later chapter we shall see why the balloon bursts with a loud bang.

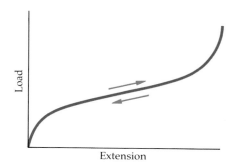

Fully elastic S-curve behavior shown my many synthetic rubbery solids. Most extension or compression occurs through a relatively narrow middle range of loads. Such solids may be brittle at higher loads.

AFTER HOOKE: THE LOGJAM IN THE EIGHTEENTH CENTURY

Seminal as these ideas of Hooke's were, they did not make much practical difference in the world of science and technology until well over a hundred years after Hooke's death. The reasons were the familiar mixture of social and conceptual ones.

Although intellectually the eighteenth century was fruitful and creative in many other ways, progress in the applied sciences was patchy. After a period of brilliant eclecticism during the seventeenth century, science in England under the influence of Newton tended to become "purer," more Greek, during the eighteenth and early nineteenth centuries. Famous applied scientists like James Watt (1736–1819) of steam engine fame were a rarity, and prestige in science attached to subjects like astronomy. The great engineers of the Industrial Revolution, such as Thomas Telford (1757–1834) who was the son of a Scottish shepherd, tended to be scornful of mathematical theory. George Stephenson (1781–1848), the "father of the locomotive," was the son of a coal miner. Stephenson could barely read or write and knew little of mathematics.

The early stages of the Industrial Revolution in England owed much to the fruitful partnership between inventive but largely uneducated craftsmen of the type idealized by Samuel Smiles in his *Lives of the Engineers* (1861) and Nonconformist bankers who put up the capital. Some of these entrepreneurial bankers were Quakers—such as the Peases, who financed the Stockton and Darlington Railway, the first important steam railroad. Others were the descendants of the Huguenots who had fled from persecution in France after the revocation of the Edict of Nantes by Louis XIV in 1685.

In contrast, the situation in France during most of the eighteenth century was almost the opposite of that in England. Applied science was encouraged, both officially and socially, and France produced a number of brilliant engineering scientists, such as Charles-Augustin de Coulomb (1736–1806). Their practical achievements tended to arise from government-inspired projects like the creation of fountains outside the royal pal-

George Stephenson's Killingworth locomotive, around 1815–1820.

I. K. Brunel standing in front of the Great Eastern.

aces and, more importantly, the design of warships. French naval architecture was well ahead of British, and the best warships in the Royal Navy in Lord Nelson's time were often captured French ones.

Unfortunately, when it came to more daring forms of industrial innovation, such as the development of steam engines and railroads, French engineers and inventors were handicapped by the weaknesses of the French entrepreneurial system. French business does not seem to have been very helpful to innovators. The Huguenot financiers had escaped to England, and French banking limped along from one scandal to another. Finally, French support for the American Revolution contributed to a fiscal crisis in the French economy.

In short, French engineering scientists seem to have had plenty of ideas but often lacked the opportunities to exploit them. The dramatic success in England of the two French emigré engineers, Sir Marc Brunel (1769–1849) and his even more famous son—who, incidentally, refused a knighthood—Isambard Kingdom Brunel (1806–1859), both of French origin and education, underscores the point.

The Fourdrinier brothers provide another example of a happy marriage between French ingenuity and British opportunity. Until the beginning of the nineteenth century all paper was made by hand in individual sheets, and it was consequently expensive and in limited supply. Henry Fourdrinier (1766–1854) and Sealy Fourdrinier (died 1847) invented and developed in England the first machine that made paper continuously, rapidly, and cheaply. They patented it in 1807. The Fourdrinier machine is the ancestor of all modern paper-making processes, and its influence—economic, political, and educational—has been immense.

FREEING THE LOGJAM: THE CONCEPTS OF STRESS AND STRAIN

After a serious conceptual barrier has been overcome and the necessary new ideas have been introduced, understood, and accepted, it is often hard to appreciate what all the fuss was about. This is true, to some extent, of the innovative engineering ideas of stress and strain. When calmly and objectively considered, their meanings are simple indeed, but throughout the eighteenth century the conceptual difficulty seems to have been real and its solution eluded many able minds. Even today, to people with no background in physics or engineering, these words of one syllable can be mysterious and alarming.

The various objects that Hooke had experimented upon and written about, such as wires, springs, and beams, were not only "specimens of material," they were also structures, each having its own individual shape and dimensions. How could one apply Hooke's results to the design or analysis of other structures made from other materials? To what extent did the behavior of a structure depend upon its material and to what extent was it controlled by its size and shape?

Throughout the eighteenth century, the handful of elasticians wrestled with the problems posed by their subject matter like Hooke did: by treating forces and deflections as characteristic of a structure as a *whole* rather than by trying first to understand the conditions existing at any given *point* within the material. Nowadays, with modern knowledge, we tend to shy away from making too sharp a division between *structure* and *material*, at least in sophisticated structures. In biology, the distinction is often not even possible. However, under the conditions at the beginning of the nineteenth century, when the objects of study were usually simple metal structures, such a distinction was a necessary and fruitful analytical procedure. Indeed, it represented an essential step forward in the history of the science of strength.

A consideration of the conditions at any chosen point within a material subject to mechanical forces gives rise to the concepts of *stress* and *strain*. The recognition of the value of this consideration as an analytical tool was one of the most important breakthroughs in the history of the study of strength and elasticity. In fact, it freed a century-old intellectual logjam.

The statement of the concepts of stress and strain in an unambiguous and usable form was due to a Frenchman. Although Augustin Cauchy (1789–1857) was, deservedly, made a baron by Napoleon III, his name is still not well known in English-speaking countries. As a boy he did brilliantly at school in both classics and mathematics, and then went on to study engineering at the Ecole des Ponts et Chaussées. After graduating first in his class, he worked from 1811 to 1813 as an engineer, garnering practical experience on the construction of the new harbor at Cherbourg. He subsequently went into academic life. His important paper on stress and strain was presented to the Academy of Sciences in 1822, when Cauchy was thirty-three. That Cauchy's ideas took some time to sink in and become widely used and accepted, especially in Anglo-Saxon countries, may have been due partly to the fact that he put them over in a generalized and mathematical way that practical engineers found difficult to understand. Furthermore, his work confronted the curious intellectual conservatism that has been so pervasive among engineers concerned with structures throughout the history of the subject. As we shall see, despite a number of recent serious accidents, there is a somewhat analogous resistance today toward using modern fracture mechanics; that is, the modern approach to the study of fracture by way of energy rather than force.

CAUCHY: STRESS

Although Kipling's sentiment was impeccable, his choice of a title was not: perhaps he meant breaking *stress;* or possibly breaking *load* or breaking force. In spite of his many stories and poems about engineers and machinery, he apparently did not know the difference between stress and strain. He had a good excuse: even fifty years ago, the careful engineering textbooks used to include tables with such heading as "Breaking Strain of Wire Ropes." What they really meant was breaking load or breaking force. An engineering student today would probably be flunked for that sort of gaffe.

Cauchy defined a stress as the load *per unit area* of the cross section at a particular point in a material. In symbols, the stress σ (sigma) exerted by a force P acting across an area A satisfies the equation

Stress (σ, the Greek letter sigma) is defined as the load (P) per unit area (A) of a cross section of material: $\sigma = P/A$. It is a condition at any particular specified point. Stress may be tensile or compressive, or, as we shall see later, it may be a shear stress. Tensile stresses are generally positive, compressive ones negative.

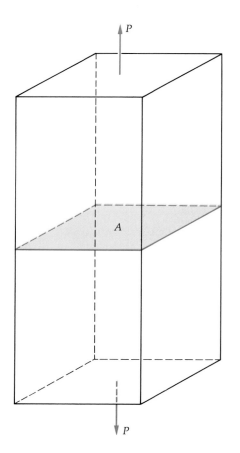

$$\sigma = P/A$$

The idea of breaking stress had been implicit in Galileo's experiments with rods in tension. As Galileo said, the breaking force or load for rods or bars of similar materials subjected to axial tension, is proportional to their cross-sectional areas. So, if a bar with a cross section of 2 square inches breaks at a load of 50 tons, a bar with a cross section of 4 square inches will break at 100 tons, and so on. Simple arithmetic brings us to the concept of a break-

Approximate Tensile Strengths for Various Common Materials

Material	p.s.i.	MN/m^2
Metals		
Steel piano wire (very brittle)	450,000	3,000
High-tensile engineering steel	225,000	1,500
Commercial mild steel	60,000	400
Wrought iron, traditional	20,000–40,000	140–280
Cast iron, traditional	10,000–20,000	70–140
Cast iron, modern	20,000–40,000	140–280
Pure cast aluminum	10,000	70
Aluminum alloys	20,000–80,000	140–550
Copper	20,000	140
Brasses	18,000–60,000	120–400
Magnesium alloys	30,000–40,000	200–280
Titanium alloys	100,000–200,000	700–1,400
Non-metals		
Wood, spruce: along grain	15,000	100
Wood, spruce: across grain	500	3
Ordinary glass	5,000–25,000	30–170
Ordinary brick	800	5
Ordinary cement	600	4
Flax fiber	100,000	700
Cotton fiber	50,000	350
Catgut	50,000	350
Spider's silk	35,000	240
Human tendon	10,000	65
Hemp rope	12,000	80
Leather	6,000	40
Human bone	20,000	140
Nylon fiber	140,000	1,000
Kevlar 29 fiber	400,000	2,700

UNITS OF STRESS

Stress can be expressed in any units of force divided by any units of area. A diverse assortment of units is in common use. Most English-speaking engineers think, and generally calculate, in pounds per square inch or in tons per square inch. Because of the confusion between American tons (2000 pounds) and British tons (2240 pounds), pounds per square inch (p.s.i. or lb/in^2) is safer. In both Western European and Eastern European countries, stress is generally expressed in kilograms per square centimeter (kgf/cm^2 or kg/cm^2).

In the International System of Units (abbreviated SI), the official unit of stress or pressure is the pascal (Pa), which is one newton per square meter. However, because this unit generally results in very large numbers, megapascals (MPa) are more usual:

$$1 \text{ MPa} = 10^6 \text{ Pa} = 10^6 \text{ N/m}^2 = 1 \text{ MN/m}^2$$

If everyone were to use SI units the world would be a tidier place. In actual practice engineers and other human beings seem to find it easier to relate loads to pounds or kilograms and cross sections to square inches or square centimeters. Megapascals, meganewtons, and square meters strike us as somehow unreal. Engineers in Eastern Europe seem to be about as resistant to pascals as engineers in the West. In England, there is a tendency to compromise by using meganewtons per square meter (MN/m^2). Americans, understandably, seem to cling to pounds per square inch (p.s.i.).

Conversion factors for units of stress

MPa -(MN/m^2)	kg/m^2	p.s.i.
1	10.2	146
0.098	1	14.2
0.00685	0.07	1

ing tensile stress: here the breaking stress is a force of 25 tons per square inch. This figure could be used for a similar rod of any thickness to predict its tensile strength (that is, the force needed to break the rod).

The concept of breaking stress seems to have been used to a very limited extent by practical engineers in the design of structures such as suspension bridges during the eighteenth century. Cauchy's innovative approach was to apply the idea of stress not just to special contexts like determining the failing strength of suspension cables, but to measure the conditions at *any* point in *any* material or structure, however far the material might be from failure. As he foresaw, in this form the concept can be a powerful intellectual tool.

The stress in a material is analogous to the pressure in a fluid, which is a much older idea. The difference is that the pressure in a fluid is exerted equally, or *hydrostatically*, in every direction, so that the resultant of the pressure of a gas or a liquid is always normal to the walls of the container whereas stress is highly directional.

STRAIN

Since Cauchy, science has deviated from the long and distinguished literary tradition of confusing strain with stress. In science, strain is the *consequence* of applying a stress to a material. A stress tells us how *hard*—with how much force—the atoms at any point in a solid are being pulled apart or pushed together. The strain tells us how *far* they are pulled or pushed; that it, by what proportion the bonds between the atoms—and the material itself—are stretched or compressed.

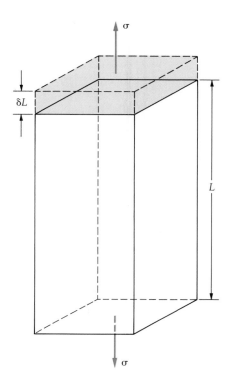

Strain, which is not to be confused with stress, specifies the effect of applying a stress to a material. Strain is defined as the ratio of the stress-induced change in length of a material to its original length. Because strain is a ratio, it has no units. Strain is usually denoted by the Greek letter ϵ (epsilon).

SHEAR

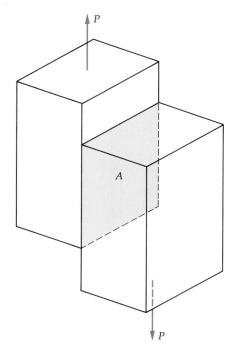

Shear stress (N) = $\dfrac{\text{shearing load (P)}}{\text{area being sheared (A)}}$

Life would be a good deal easier and simpler for engineers and elasticians if the loads, stresses, and strains to which materials and structures are subjected were all tensile or compressive ones. However, besides tension and compression, we frequently have to deal with shear. If tension is associated with pulling and compression with pushing, then shear is associated with sliding or, more specifically, with resisting a tendency to slide when a force is applied. Fortunately, the concepts of shear, shear stress, shear strain, and shear modulus (or "stiffness" in shear) are analogous to their equivalents in tension and compression.

Shear stress τ (tau) is the shear force per unit area of the cross section of the material on which the force is acting:

$$\tau = \frac{\text{shearing force}}{\text{area of cross section on which force acts}}$$

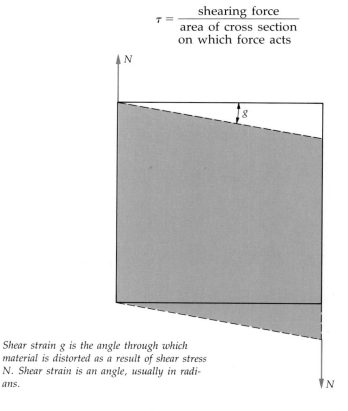

Shear strain g is the angle through which material is distorted as a result of shear stress N. Shear strain is an angle, usually in radians.

The units of shear stress are the same as those of tensile and compressive stress: megapascals (MPa), meganewtons/m^2 (MN/m^2), kg/cm^2, or p.s.i.

Shear strain γ (gamma) is an angle, usually given in radians.

The shear modulus G is the stress that would distort an element of the material though an angle equal to one radian. The units of G are those of a shear stress: MPa, MN/m^2, kg/cm^2, or p.s.i.

There is a simple relationship between the Young's modulus E of a material and its shear modulus G:

$$G = \frac{E}{2(1 + \nu)}$$

where ν is Poisson's ratio.

Although the need for shear strength and shear stiffness gives a great deal of trouble to engineers, especially in the design of aircraft, Nature appears to be very clever in evading shear requirements in the structures of plants and animals, as we shall see.

Strain is usually denoted by the symbol ϵ (epsilon): Thus, if a rod of original length L is caused to stretch by an amount l, it is subject to a strain ϵ as follows:

$$\epsilon = l/L$$

Strain is thus a measure of the extent of change in the length of a material due to stress-induced extension or compression.

Because strain is the *ratio* of one length to another, it is a fraction with no units. In engineering materials strain is often a very small number, typically about 10^{-3} or less, and it is commonly expressed as a percentage: e.g., 0.1 percent. Although the elastic strains in ordinary engineering materials are usually below 1.0 percent, soft materials like rubber can often extend elastically to about 800 percent, and some biological materials can stretch even farther.

A frog demonstrates the elasticity of biological materials.

YOUNG: THE STIFFNESS OF MATERIALS

As we have mentioned, one of the most important intellectual impediments in the study of structures throughout the eighteenth century was the fact that it was not possible to calculate how much of the deflection of

Thomas Young (1773–1829)

a structure under a given load was due to the geometry of the structure itself, and how much was due to the intrinsic stiffness of the material from which it was made. The idea that each type of solid has a characteristic stiffness of its own—steel having one stiffness, wood another, and so on—is the legacy of the British scientist Thomas Young (1773–1829). Young's work showed that overall deflection of a structure when it has to resist a given load is due to the *combined* effects of the stiffness of its material and of the size and shape of the structure.

Young, born to a Quaker family in modest circumstances, earned his living from the age of fourteen to nineteen by working as a private tutor in a wealthy household. His employers must have allowed him ample time for his own studies, for he taught himself not only Latin, Greek, Persian, French, Italian, Hebrew, and Arabic, but also philosophy and mathematics. When he was nineteen he found the means to study medicine, which he did successively at three universities. He set up in practice in London in 1799. However, he had already been made a Fellow of the Royal Society in 1794, when he was only twenty-one, in recognition of a paper he had written on the accommodating power of the human eye.

In the midst of activities that included his medical practice, a teaching post, various official positions, and a hobby deciphering Egyptian hieroglyphics, Young found time to publish, in 1807, his ideas about the intrinsic stiffness of materials—the concept we now call Young's modulus. Unfortunately, Young's industry and scientific originality were not matched by his power of communication. It was said of him by a contemporary that "his words were not those in familiar use, and the arrangement of his ideas seldom the same as those he conversed with. He was, therefore, worse calculated than any man I ever knew for the communication of knowledge." Not surprisingly, he was a bad lecturer.

Young's own definition of his *modulus* (Latin for small measure) reads as follows:

> The modulus of the elasticity of any substance is a column of the same substance, capable of producing a pressure on its base which is to the weight causing a certain degree of compression as the length of the substance is to the diminution of its length.

Naturally, few people at the time, and not many since, had any idea what, if anything, this meant.

However, the fault was not entirely Young's. His idea is, in fact, very difficult to express without using the language of stress and strain. If we read Young's definition carefully, we can see that he was trying to define what we now call a *specific modulus,* that is, the stiffness of a material in proportion to its density.

The idea of Young's modulus was first put into its modern mathematical form, using Cauchy's concepts of stress and strain, in 1826 by another French engineer, Claude-Louis-Marie-Henri Navier (1785–1836). Navier's formulation was

$$\text{Young's modulus } (E) = \frac{\text{stress } (\sigma)}{\text{strain } (\epsilon)}$$

In other words, Young's modulus is the slope of the stress-strain curve: it compares the size of the imposed stress with the size of the resulting strain. A relatively small Young's modulus would indicate that the material requires a comparatively small amount of stress to achieve each unit of strain: the material is flexible. A large modulus would indicate that the material requires a large amount of stress to achieve each unit of strain: the material is stiff.

Because this ratio is composed of a stress divided by a unitless number or ratio (strain), Young's modulus is also a stress. Its units are pascals (Pa), pounds per square inch (p.s.i.), kilograms per square centimeter (kg/cm^2), meganewtons per square centimeter (MN/cm^2), or whatever.

Young's modulus happens to be equal to the stress that would, in theory, double the length of the specimen—that is, if it didn't break first. For this reason the modulus of any reasonably stiff material is likely to be a large number. Some characteristic Young's moduli are shown in the table below. Note that the range of values is enormous: from about 170,000,000 p.s.i. (1,200,000 MN/m^2) for diamond to 30 p.s.i. (0.2 MN/m^2) for very soft biological tissues—a range of nearly 6 million to 1.

The slope E of the straight part of the stress-strain curve is Young's modulus, which has a characteristic value for each different material.

Approximate Young's Moduli of Various Solids

Solid	E in p.s.i. $\times 10^6$	E in MN/m^2
Soft biological tissues	0.00003	0.2
Rubber	0.001	7
Wood (spruce)	1.6	11,000
Ordinary cement	2.5	17,000
Bone	3.0	21,000
Ordinary glass	10.0	70,000
Magnesium metal	6.0	42,000
Aluminum alloys	10.5	73,000
Steels	30.0	210,000
Diamond	170.0	1,200,000
Nylon fiber	0.8	5,500
Kevlar 29 fiber	9.0	62,000
Kevlar 49 fiber	19.0	130,000

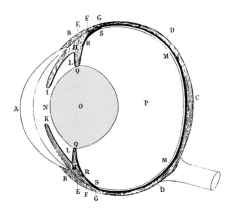

Thomas Young drew attention to the focussing mechanism of the human eye in 1793. This is an important application of the concept of Poisson's ratio.

YOUNG AND POISSON: STRAIN IN THREE DIMENSIONS

So far we have considered the effect of applying a stress to a material as being one-dimensional. In other words, the material will be strained in the direction of the applied stress according to the formula

$$\text{Strain} = \frac{\text{stress}}{\text{Young's modulus}}$$

or, in symbols,

$$\epsilon = \frac{\sigma}{E}$$

This description is true as far as it goes. However, it takes no account of what this stress does to the material in the other two dimensions—that is in the directions normal to the axis of the applied stress. If one pulls on, say, a rubber band, it is easy to see that the rubber becomes longer but thinner. The material not only *stretches* in the direction of the pull; it *contracts* in the other two dimensions.

As mentioned earlier, Thomas Young was made a Fellow of the Royal Society at the improbable age of twenty-one for his 1793 paper on the accommodating power of the human eye. In this paper he pointed out that the lens of the eye is made of a transparent, elastic, and fairly flexible material. A muscle encircling the lens enables it to change its focal length

When a solid is stretched by a tensile stress σ_1, it undergoes a primary strain ϵ_1 in the direction of σ_1, but it contracts laterally by a secondary strain ϵ_2. Poisson's ratio ν is equal to ϵ_2/ϵ_1.

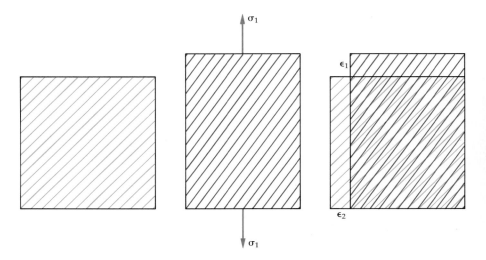

so that the eye can focus on either near or distant objects. When this muscle is relaxed, the lens is at its maximum diameter and at its minimum thickness. The radius of curvature of the surfaces of the lens is therefore at its maximum, and so is the focal length. A person with normal eyesight can then focus at objects at infinity. To focus on nearer objects, we contract the muscle encircling the lens. This causes the elastic lens to contract in diameter but at the same time to bulge in the middle so as to become thicker and more sharply curved. This, of course, reduces the focal length and enables you to read this book.

Important as this observation was, Young was content to describe the strains induced in the lens at right angles to the applied stress in a generalized, qualitative way. Not until 1829 (the year of Young's death) was the matter stated in a precise, mathematical, and practically usable form. Again, a Frenchman was responsible.

Siméon-Denis Poisson (1781–1840) was another hard-working French scientist who rose from humble beginnings to achieve distinction in the realm of mathematics. His fame as a mathematician led to his induction into the French Academy in 1812. He took up the idea of using the gross elastic behavior of a solid as a means of investigating its fine-scale molecular structure—in other words, as a way of finding out about the behavior of interatomic bonds. To this end Poisson considered elastic strains, not in one dimension, but in three.

When a Hookean, elastic material is subjected to a stress σ_1 and not otherwise constrained) along one axis, it will, of course, show a strain ϵ_1 along that axis:

$$\epsilon_1 = \frac{\sigma_1}{E}$$

Recall that E stands for Young's modulus. Poisson showed that the stress will also cause strains ϵ_2 and ϵ_3 along the two axes normal to the σ_1 axis, and that these secondary strains will relate to the primary stress and strain as follows:

$$\epsilon_2 = \epsilon_3 = -\nu\epsilon_1 = -\nu\frac{\sigma_1}{E}$$

The proportionality constant ν (nu) is known as Poisson's ratio. Note that, the *sign* of ϵ_2 and ϵ_3 must be opposite to that of ϵ_1 as indicated above. Strictly speaking, ν should therefore carry a minus sign, but in practice the minus is generally omitted.

Poisson's formula indicates that if the primary stress and strain are tensile, the secondary strains will be compressive; if the primary stress and strain are compressive, the secondary strains will be tensile. For most ordinary engineering materials, Poisson's ratio has a value between one-fourth

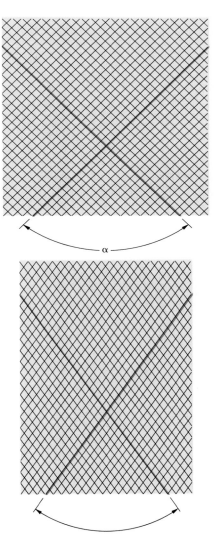

Anisotropy and Poisson's ratio. The smaller the angle α the higher the Poisson's ratio and the higher the anisotropy. Isotropic materials mostly have Poisson's ratio equal to about $\frac{1}{4}$ to $\frac{1}{3}$.

STRAIN ENERGY: THE ENERGY OF A SPRING

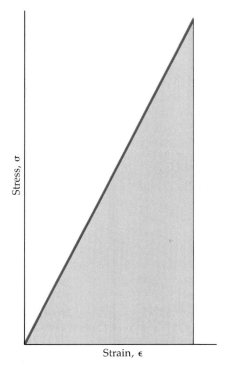

Strain energy η is equal to ½σε, the area under the stress-strain curve.

When an elastic material is strained, energy is stored in it. This energy is released when the strain in the material is relaxed. A lot of energy may be involved. The strain energy stored in bows and catapults has been utilized by warriors for thousands of years. The vibration of a violin string, or a tuning fork, or a bell is due to the rhythmic interchange of strain energy with kinetic energy in these devices.

The energy η (eta) stored in an elastic, Hookean material that is subject to an axial stress σ causing a strain ϵ is given by

$$\eta = \tfrac{1}{2}\epsilon\sigma$$

Strain energy was traditionally expressed in foot-pounds, but nowadays it is generally more convenient to use joules. One joule is the energy expended when one newton acts through one meter.

The first person to make use of the concept of strain energy in engineering analysis seems to have been Leonhard Euler (1707–1783), who was handicapped since the ideas of stress and strain had not yet been articulated.

The idea of strain energy as an engineering tool is an old one. Its utility in the past tended to be confined to analyses of springs, vibrations, and the buckling habits of panels and columns under compression. The wider use of strain energy in the study of the tensile fracture of all kinds of structures is comparatively new, and not always welcome to traditional engineers. The concept is, however, useful in considering problems of safety, and we shall return to the subject in Chapter 4.

and one-third—typically about 0.3. It is easily shown that, as long as Poisson's ratio is less than 0.5 the material will expand in volume when it is under tension and contract when it is under compression. The larger the value of Poisson's ratio, the more fully the secondary strains are compensating for the primary strain, and the smaller the overall change in the volume of the material. When Poisson's ratio is 0.5 there is theoretically no volume change. This is the case with liquids which cannot resist axial stress; and it is also very nearly true of some jellylike substances, such as the lens of the eye, with which Young was concerned.

In a homogeneous, *isotropic,* material (that is, one having the same properties in all directions—a metal for instance) Poisson's ratio cannot be greater than 0.5. A higher number would indicate an overall *gain* in energy during the process of stress and strain—a situation that, as Euclid might have said, is not only impossible but absurd. However, many biological materials have a very complicated molecular morphology that can make them anisotropic (that is, having with such materials, different elastic properties in different directions). For example, as one can see by feeling the muscles in one's arm, which bulge considerably when contracted, Poisson's ratio may be much larger than 0.5; it is often around 1.0.

Young's 1793 paper in which he applied engineering concepts to a biological problem—the focusing of the mechanism of the eye—and anticipated the idea of Poisson's ratio was pathbreaking in its interdisciplinary approach. Young may rank as one of the patron saints of the modern science of biomechanics. He was fortunate in having his work recognized as not only respectable but important by the Royal Society and the scientific establishment of the day; this might not have been the case a few years later, when once again conservatism prevailed in academic institutions and scientists tended to work within the confines of a single field of study. Furthermore, just as Young did not hesitate to apply his mechanical ideas to a medical problem so he did not let his position as a doctor prevent him from making a very important contribution to the study of masonry walls and arches and dams.

After Young's time the barriers tended to go up. Even in the 1930s, when I was an engineering student at the University of Glasgow, a school with considerable but separate traditions in medicine and engineering, I remember being scorned by medical students, my contemporaries, who made remarks such as "all that stuff about Young's modulus!" Perhaps they did not know Young was a doctor. Indeed, until quite recently, scientists with interdisciplinary interests were apt to endanger their careers by being regarded as "unsound." The modern fashion for biomechanics is surprisingly recent. If there is such a thing, the curve of scientific progress is not a smooth one. It is more nearly a series of steps that are initiated or delayed by the intellectual conventions—not always strictly rational—of the age. We may have considerable respect for Thomas Young, however incomprehensible he may have been.

3

THE FORM AND DIVERSITY OF STRUCTURES

The Weight, the Cost, and the Purpose

*Here's my wisdom for your use, as I learned it when the moose
And the reindeer roamed where Paris roars to-night:—
"There are nine and sixty ways of constructing tribal lays,
"And—every—single—one—of—them—is—right!"*

RUDYARD KIPLING

Nearly every structure exists for some purpose, although that purpose may not always be easy to understand—especially if the structure is a biological one. Many structures fulfill more than one purpose, either simultaneously or successively. A common cause of accidents to engineering structures is the designer's failure to anticipate correctly the magnitude and direction of the loads that will have to be resisted. But loads are only one of the sources of ambiguity in engineering.

In the "purer" sciences such as astronomy and chemistry, many problems have only one correct answer: problems like calculating the orbit of Venus or the number of electrons in a chlorine atom. For the engineer, on the other hand, there is often more than one way to do a job. Loads may be supported in tension or compression—by pulling or pushing them. The designer may be able to choose between a suspension bridge or an arch, a big tent or a cathedral. One of these solutions is not necessarily better than another. Bridges, for example, exist in considerable variety, as any visitor to Manhattan will have noticed.

In structural technology similar effects may be produced by using opposing methods. The bridge over the Rhine at Reichenau, Switzerland, is an arch, a structure that supports its loads chiefly in compression. The Brooklyn Bridge is a suspension bridge, a structure that supports its loads mostly in tension. Both achieve the same result.

If choices abound in engineering, they are even more profuse in biology. Faced with the problem of evolving a small flying animal, Nature has produced birds, bats, and butterflies. Although these creatures exemplify three widely differing structural "philosophies," they are all quite viable and manage to coexist in a highly competitive world. They may well be nearly equally efficient as structures.

The availability of multiple solutions to a given structural problem is one of the reasons why plants and animals have evolved into such an enormous variety of forms. Similarly, the existence of equivalent structural alternatives explains why the potential for development in technology is so great. The task of designing or developing an efficient structure is no easy one, however. In biology, millions of years of natural selection have provided an encyclopedic variety of imperfect or intermediate structures; in technology, too, engineers have had to learn the hard way. The moral is that we can never afford to be dogmatic or narrow-minded. The doors are always open for unconventional approaches to traditional problems.

Nature has evolved many solutions to the problems of designing a small flying animal. Although the materials used and the structural features vary widely from bird to bat to insect, the results are almost equally successful.

THE CHOICES OF MATERIALS

Before we proceed to describe and compare some of the solutions to traditional problems in biology and engineering, we will consider the question of how to distinguish a *material* from a *structure*. As we have mentioned in Chapter 1, the distinction between a structure and a material is often an artificial one. Engineers have been accustomed to build their structures out of a limited range of materials—a few metals, concrete, and so on—and to regard the study of these substances as a subject separate from *real* engineering. But Nature makes no such distinction. Every animal is an elaborate hierarchy of structures, substructures, and subsubstructures. It would be difficult to identify the material or materials from which, say, human beings are made. We can say that trees are made largely from wood, which is a readily identifiable substance, but wood itself is an elaborate and sophisticated structure, as we shall see.

Even the metals engineers use are not simple uniform substances; metals have complex and variable crystalline structures that have only been understood recently. And technology is developing materials with increasingly convoluted structures: for example, artificial fiber-resin composites, which are conceived on biological models. The more advanced forms of engineering seem to be mapping out an anatomical progression in which it is increasingly difficult to draw a clear dividing line between the structure itself and the materials from which it is made.

Despite the literal deficiencies of the term, it is still convenient to talk about a material. As long as we are aware that we are using the word loosely, we can indulge in such truisms as "no structure can be better than the materials from which it is made." But what is a "better" material? The layperson might suppose that a better material is simply one that is stronger, cheaper, and more durable. By strength most people mean tensile strength. The problem with this quick definition is that wooden structures, for instance, seldom fail to support loads in tension; nylon is very strong in tension but its structural applications are limited. Even when we have defined what we really mean by *strength,* we may find that strength, by itself, is not enough. We have to consider a number of other physical properties of materials before we can begin to predict the strength of a structure.

Even when we have determined what we think we need in a material for a given structure, we are likely to find that there is no unique answer to the question "what is the best material to use?" Not only may numerous different forms prove to be nearly equally effective, but we often have a surprisingly wide choice of materials. During World War II, aircraft built from wood often turned out about the same weight and cost as similar aircraft built from light alloys, and performed as well. Nowadays, yachts are built from wood, steel, aluminum, and reinforced plastics. These materials give approximately equal results in terms of weight, cost, and reliability in service.

All this does not mean that the choice of materials is unimportant. As in the construction of Kipling's tribal lays, there may be nine and sixty ways of achieving a good result, but the choice of words matters greatly to the poet. The analogy with matrimony is tempting. There may, perhaps, be nine and sixty people in the world to whom one might be happily married, but, though each of these hypothetical marriages would be different, one would not necessarily be better than another. But of course, there are far more than sixty-nine people in the world with whom marriage would be, in varying degrees, disastrous. The choice of words, marriage partners, and materials, though fairly wide, is important.

We mentioned that the engineer's choice of materials has been getting wider every day. When we come to exploit the mechanical properties of this extensive range of materials by using them to make structures, however, we find that the choice of structural form is not as wide as the choice of materials. Although there is nearly always more than one way of achieving a given structural purpose, the morphologies or shapes of structures do tend to fall into a fairly small number of catagories and archetypes.

Although the materials found in biology are often very different from those used in engineering, the geometries of the structures in which materials can be employed to carry loads are generally much the same. Of course, within any given structural category there is room for a great deal of variation and ingenuity in the detailed design. Nature is frequently more clever than engineers at developing the potential of a given structural concept.

One way of looking at the science of strength is to consider the evolution of structural forms that are most suited to exploit the strength properties of materials. Another approach is to assume that both natural and synthetic materials have evolved in such a way that they are most effective when used in a structure. A "strong" material that is not adapted for use in a structure has little or no place in the scheme of things. So, if we want to understand materials science, we must first know something about the shapes and functions of some basic kinds of structures.

TENSION VERSUS COMPRESSION AND EULER BUCKLING

As we said at the beginning of this chapter, a designer is sometimes able to choose between building a structure in which the loads are supported mainly in tension and one in which they are carried mainly in compression. Usually some combination of the two is desirable. In any case, the designer typically has a certain amount of choice in the matter.

Sometimes supporting a load in tension presents more problems and incurs more risks than supporting the same load in compression; but this is by no means always true. With unreliable materials and with primitive joints, for example, tension is best avoided; but in biology as well as in modern technology, a tension structure is often the lightest, cheapest, and safest solution. Traditional compressive constructions, such as masonry and log huts, have the advantage of being simple and fairly reliable, but they are heavy and consume a lot of labor. Nature does not offer many examples of compressive structures; anthills are a rare exception. Such a design is quite unsuitable for structures that are mobile or lightweight—characteristics associated with living things.

A thin rod, rope, or membrane under tension will eventually tear. Its strength is not affected by its length under static loading.

A short strut is likely to fail by local crushing—that is, by the actual fracture of the material.

A longer rod, strut, or membrane will not fail by fracturing, but by bending or buckling. This is likely to occur at a small load.

The behavior of a structure in compression depends a great deal upon how far the load has to be carried, that is, upon what is called the structure loading coefficient—which may be defined for now as the ratio of the magnitude of the load to the distance over which it has to be transmitted.

The way in which materials fail under tension differs from the way they fail under compression. If one pulls on a tension support or *member*—whether it be a rope, a steel rod, or an animal tendon—it merely gets straighter the harder one pulls. Failure in tension usually occurs by a separation of the molecules at the weakest cross section of the member. With a uniform member, the length does not make any difference to the strength in tension under static loading.

The behavior of materials that fail under compression depends on their length. A low wall of bricks, a short metal strut, or a short bone may remain straight and stable under the load until the material finally succumbs to a local crushing mechanism. A longer strut, a higher wall, or a yacht's mast are liable to buckle—that is to bend or bow—even under a light load.

A simple experiment with an ordinary piece of paper serves to illustrate the enormous difference between a long, thin structural member in tension and a similar member in compression. If one first attempts to pull the paper apart in its own plane, and then attempts to push opposite edges of the paper closer together in the same plane, one will observe an effect that is of the greatest importance in both biology and engineering.

The Swiss mathematician Leonard Euler (1707–1783)—pronounced "Oiler"— devised a formula for calculating the load at which a long rod will buckle when it is subjected to a compressive force along its length. Euler's human—indeed his practical—interests seem to have been distinctly limited, but he was an almost prototypal example of a mathematical genius. He obtained his master's degree in mathematics at the age of 16, and published his first research paper when he was 20.

Euler was apparently not much interested in the buckling strength of long columns from a technological point of view. His interest in the problem arose rather because he had just invented the calculus of variations, and he was looking for a problem to try it out on. Someone suggested that he might attempt to calculate the height of a vertical rod that would buckle under its own weight—a sort of mathematical investigation of the Indian rope trick. Euler succeeded in calculating the length of the rod, and his first results were incorporated in a book published in 1744.

Although Euler's calculations are correct as far as they go, his work antedated the concepts of stress and strain by almost a century, and he did not have access to Young's modulus. For these reasons his work would have had limited practical application at the time—even if it had been understood by his contemporaries.

CHAPTER THREE

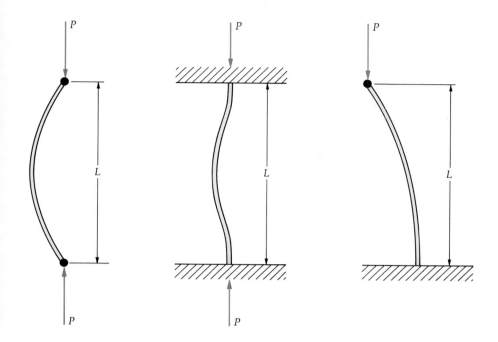

Various Euler conditions. A strut or panel that is free to hinge (left) has a buckling load P equal to $\pi^2(EI/L^2)$, where E is Young's modulus, I is the second moment of area (see page 00), and L is the length of the strut or panel. If both ends are fixed in position and direction, the buckling load is equal to $4\pi^2(EI/L^2)$. With one end encastré and the other pin-jointed and free to move sideways, P is $\pi^2(EI/4L^2)$.

In the modern language of stress, Euler says that a long rod or column will buckle at an axial load P according to the equation

$$P = k\pi^2\frac{EI}{L^2}$$

Where E is Young's modulus, I is the second moment of area of the cross section of the rod (sometimes wrongly called the moment of inertia),L is the the length of the rod, and k is a constant that depends on the end conditions—that is, on the extent to which the ends of the rod are held stationary or are free to rotate. When both ends are pin-joined (i.e., allowed to rotate), the value of k is 1. When both ends are clamped (i.e., fixed in direction and position), the value of k is 4. When one end is fixed and the other free, k is equal to ¼.

Hence the buckling load depends on:

1. *The square of the unsupported length of the rod.* In other words, the shorter the rod, the stronger it is. This very important factor has an enormous influence on the shape and weight of both man-made and living structures.

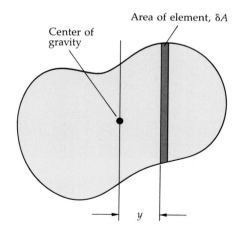

The second moment of area, I, sometimes called the moment of inertia, is expressed as ∫ δA(Y²).

2. *Stiffness*. A high Young's modulus is essential. A steel, aluminum, or wooden rod has a much higher axial compressive strength that one of the same cross selection made from, say, nylon or tendon. Note that stiffness or axial compressive strength cannot be generalized to other kinds of strength: the tensile strength of nylon is roughly the same as that of steel. The importance of the E value explains why so much money has been spent on the development of high-stiffness fibers—boron and graphite—for advanced composite materials.

3. *The second moment of area of the cross section of the rod*. Because the range of high-stiffness materials is somewhat limited, an easier and usually cheaper way to increase the buckling load of a strut is by changing the geometry of its cross section. The higher the second moment of area, the stronger the rod. The I value can be raised by using hollow sections, such as tubes, as struts, or by fashioning I-section metal joists. In biology, cellular tissues and hollow bamboos and leg bones utilize this geometrical strategy to maximize their resistance to buckling. In other words, a section with some space in or around it is generally a good thing.

The *buckling* load of a strut or a panel is not necessarily the breaking load. In fact, as we can easily deduce from Euler's formula, the actual compressive stress at which a long strut will buckle may be almost ridiculously small. So, when the compressive load is taken off, the strut may recover undamaged. It may simply spring back like a bow whose string has been cut.

In the real world this is quite a good way of dealing with adversity. For instance, carpets and door mats utilize this strategy: when someone treads on a carpet, the pile buckles and then recovers. More significantly, Euler buckling provides an important botanical safety mechanism. Small plants such as grasses and very young saplings cannot possibly be made strong enough to stand up rigidly when animals and people trample on them. Why is grass so hardy? The answer is, in part, because the stems are long and thin so that they can buckle when they are stepped on. The buckling stress is so small that they do not usually fracture but are able to recover, either immediately or after a short interval. This recoverability is what makes lawns and grass playing fields practicable. When a plant gets too big to be trampled on, it is generally able to stand up more stiffly against the wind because the stem or trunk has become much thicker in proportion to its height.

Euler buckling as a safety mechanism does not often appear in animals. However, my colleague, Dr. Julian Vincent, recently discovered an interesting example of this phenomenon in the spines of the common hedge-

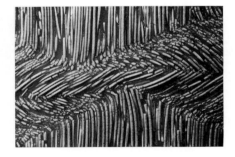

100 µm

Buckling of aluminum oxide under longitudinal compression.

The common hedgehog (Erinaceus). The sharp spines of the hedgehog's back serve to protect it from predators such as foxes, dogs, and children, but they also have another function. The animal habitually descends from trees and other high places simply by rolling or falling. When the rolled-up beast hits the ground, its spines buckle in an Eulerian manner and so cushion the impact.

hog. The hedgehog is a small animal common in Europe, Asia, and Africa. Its body is covered with sharp, centimeter-long spines set close together. When the animal is alarmed, it curls itself up into a spiky ball that is more or less proof against dogs, children, and other predators. Until lately it was supposed that this defensive role was the sole purpose of the spines. Surprisingly, for so apparently clumsy an animal, hedgehogs are often arboreal. They do climb trees, possibly in search of insects. Although they can climb up, they are sometimes unable to climb down again. Their solution to the problem of descent is to curl themselves into a ball and simply roll or fall to the ground. On impact with the ground, the spines buckle in an Eulerian manner and thus absorb the energy of the fall and mitigate the shock of impact. In fact, Vincent shows that the design of the spines seems to be optimized for this purpose.

Further study may well reveal applications for Euler buckling as a safety mechanism in technology. However, engineers are not at present dedicating themselves to utilizing Euler buckling in this fashion. They are generally preoccupied with the problem of preventing a structure from buckling under the loads it is likely to experience in service.

THE RELATIVE WEIGHTS OF SIMPLE
TENSION AND COMPRESSION STRUCTURES

The efficiency of tension and compression members can be compared by determining the weight of the lightest member that is able to support a given load in each case. This figure is the *weight-cost* of sustaining the load. Because the strength of a tension member is roughly independent of its length, whereas that of a compression member (as estimated by the Euler buckling load) diminishes as $1/L^2$, the relative weight-cost of sustaining a given load in tension or in compression can vary enormously. The weight-cost also varies according to the length over which the load has to be supported. When the loads are relatively large and the distances are short—as within an automobile engine—there may not be much to choose between tension and compression. When the structure become more diffuse, as the distances become greater and the loads relatively lighter, the difference in the weight of the supporting member and often in monetary cost can be very significant indeed.

Consider, for instance, the weight-costs of carrying 1,000 kg (10,000 newtons) over 10 centimeters, 1 meter, and 10 meters, in tension and in compression. The rough figures in the table below are based on the properties of ordinary engineering steels. The compression members are assumed to be steel tubes with cross sections of the strongest possible design; in the tension members the geometry of the cross section does not make much difference. A reasonable allowance has been made for the weight of the end fittings that would be needed to get the load into and out of the member in each case.

The table shows that over short distances compression members may be lighter than tension ones; this is mainly because compression members need much less in the way of end fittings. When the distance becomes

The weight-cost of carrying one metric ton (1,000 kg) over various distances in tension and in compression

	Distance over which load is carried		
	10 cm (4 in)	1 m (3.3 ft)	10 m (33 ft)
Weight of member in tension	1.8 kg (4.0 lb)	1.93 kg (4.2 lb)	3.5 kg (8.0 lb)
Weight of member in compression	0.2 kg (0.5 lb)	2.0 kg (4.5 lb)	200 kg (450 lb)

greater, the compression members become 10 or 100 times heavier. And, of course, the monetary cost of an engineering structure or the cost to a plant or an animal in terms of metabolic energy may be proportionately higher for the longer distances.

The efficiency of a compression structure can be quantified by a simple ratio. The *structure loading coefficient* may be defined as the ratio of the size of a load carried by a member to the distance the load has to be carried:

$$\text{Structure loading coefficient} = \frac{\text{load}}{\text{distance load must be carried}}$$

The higher the structure loading coefficient, the more efficient the member. Efficiency increases when the design of the overall structure is such that either the permissible load increases or the distance it has to travel is shortened.

These considerations have an important influence on design in mechanical engineering. In the construction of heat engines, for instance, as we have increased the working pressures and the rate of rotation, we have increased the load on the parts. At the same time we have been able to reduce their dimensions substantially. In order words, we have increased the structure loading coefficients in a very beneficial way. The practical effect can be appreciated by comparing the weight and cost of a modern internal combustion engine with that of an early steam engine of equivalent power. One of James Watt's 10-horsepower beam engines might have weight over 10 tons. A modern 10-horsepower gasoline engine in a motorcycle or an outboard motor might not weigh much more than 10 pounds.

A rather interesting example of the influence of structure loading coefficients in technology occurs in the design of road vehicles. In a pedal bicycle the wheels are relatively large in diameter and they are lightly loaded, making it possible to save much weight by using wire spokes in tension. In fact, the bicycle is primarily a tension structure. Not only are the wheel spokes and the tires in tension, but the power is transmitted by means of a chain drive from an "engine" whose main members, the rider's muscles and tendons, are also in tension. In an automobile the loads are much higher; the wheel spokes (or the equivalent structure) are largely in compression; and the engine and transmission are mostly in compression or shear. In most cars, almost the only parts that are in tension are the tires.

Increasing the structure loading coefficient affords a convenient and welcome way out of Euler-type difficulties for traditional mechanical engineers, but this strategy may not be so readily available to structural engineers. The width of a river that has to be bridged and the magnitude of the loads that have to be carried across it may be difficult to alter.

THE EFFECTS OF SUBDIVIDING A STRUCTURE

Four thin load-bearing members—in this instance the cables supporting a suspension bridge—work as well as one thick support.

It is frequently expedient to arrange for a load to be shared between several members. The body of a parachute jumper hangs by many cords from the parachute; most tables, and many animals, are supported by four legs; and so on. The effect of this subdivision upon the weight, the cost, and the safety of a structure as a whole is quite different in compression than in tension.

If we subdivide a compressive load in such a way that it is not carried by a single strut but is shared equally between n struts, then the total weight of the supporting structure increases by a factor of $\sqrt{n}$. For example, four legs, struts, or columns are twice as heavy as one, for a given service. It follows that when humans and birds evolved to stand and walk upon two legs so as to free their front limbs for arms and wings, the result was a gain in structural efficiency.

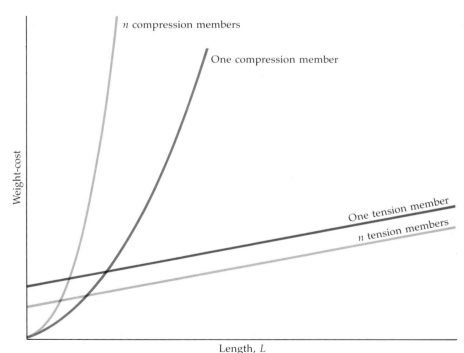

The relative weight-cost of carrying a given load over a distance L.

When we consider the case of compressive loads carried in panels—a common feature of ships and aircraft—the effects of subdivision are even more unfavorable. Other things being equal, the weight of the system increases by a factor of $n^{2/3}$.

In tension, however, the situation is different. Because tension members are not liable to buckle, it does not generally matter how thin they are. Hence, if we subdivide a rope, wire, or panel, the tensile strength of the system depends on the *total* cross section. For example, two ropes, each of 1 square inch in cross section, are as strong as one rope with a cross section of 2 square inches. Furthermore, by a more drastic subdivision we may well increase the safety of the system, because if one member breaks, the remaining ones may still be able to carry the load.

If we look more closely at actual subdivision situations, we find that loads may actually decrease the weight of the supporting structure. This is because the aggregate weight of the end fittings needed to get the load on and off of the members is slightly reduced.

With animals, structure loading coefficients are an inappropriate measure of metabolic or structural efficiency. The pressure of the various working fluids and the rates of the various metabolic reactions cannot be raised very much; animals cannot, to any extent, follow the example of the internal combustion engine. Consequently, their structure loading coefficients remain low by the standards of the automobile industry. The emphasis in biological structures is less on brute power and more on economy of means; bones are as thin and light as possible, and most other supports are tension members such as tendons, muscles, skin, and other membranes.

LIVING WITH COMPRESSIVE LOADS

Although in lightly loaded structures—that is in structures where the structure loading coefficients are low—compression loads may be more difficult to deal with efficiently than tensile ones, this state of affairs is, in practice, very common both in biological and man-made structures. It is therefore necessary to find means of coping with it without incurring excess weight, expense, and general clumsiness.

There are really three ways in which Euler buckling can be postponed or prevented when the structure loading coefficient is low:

Biological materials, such as skin, may be prestressed. As people get older, their skin tends to lose its prestress and wrinkle under compression.

1. By prestressing the material, that is, by putting the member which is liable to buckle into initial tension—rather like the skin membrane of a child's balloon. Pneumatic tires are an obvious example; so are sandbags. This sort of mechanism is very common in plants and animals. Many soft plants can hold themselves erect because the cell walls are prevented from buckling by virtue of the hydrostatic pressure of the sap which prestresses the cell membranes something like the skin of a rubber balloon. But the membrane does not *have* to be soft. As a plant matures its cell walls often harden but the hydrostatic pressure still assists them to resist buckling and thus helps the plant to stand up. There are many examples of prestressing in animals. The skin of young people is a case in point; as we get older our skin tends to lose its prestress and so wrinkle when under compression. Engineers sometimes to make use of this principle by putting panels which are liable to buckle into an initial tension by one means or another.

2. By changing the geometry of the cross section of the parts which have to resist compression—that is, by increasing their second moment of area. Solid structures can be replaced by hollow tubes—to follow the design of the bamboo or, indeed, that of many bones. Engineers often use tubes or H-section members. Sheet materials are often corrugated, as with sheet-metal or cardboard.

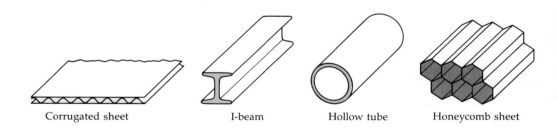

Corrugated sheet   I-beam   Hollow tube   Honeycomb sheet

3. By changing the material. Some materials are much better than others at resisting compression in circumstances where the structure loading coefficients are low.

Let us consider what makes a "good" material in compression.

Recall that the formula for the Euler buckling load of a strut of length L fitted to carry a load P is

$$P = k\pi^2 \frac{EI}{L^2}$$

Manipulating this formula algebraically, we can easily show that the weight W of this strut is

$$W \propto \frac{\rho\, L^2\, \sqrt{P}}{\sqrt{E}}$$

where E is the Young's modulus of the material and ρ (rho) is its density.

In choosing a material to make a strut that has to carry a compressive load, the crucial consideration is the value of $\rho/\sqrt{E}$. Usually it is more convenient to invert this ratio to $\sqrt{E}/\rho$, which is called the *material efficiency criterion*. The higher the value of $\sqrt{E}/\rho$, the more efficient the material is in compression.

The $\sqrt{E}/\rho$ ratio is applicable to struts and columns—that is, to compression members that can buckle in either of two dimensions. However, many compression members take the form of panels, which are able to buckle in only one dimension—that is, sideways. For panels, the mathematics is slightly different and results in a material efficiency criterion of

$$\frac{\sqrt[3]{E}}{\rho}$$

The values of the two material efficiency criteria can be used to rate the comparative efficiencies of various natural and synthetic materials in resisting buckling loads. The table below sheds light on choices made both in technology and in nature. As a material for making panels, aluminum is

Efficiency of compression structures of various materials in resisting Euler buckling

Material	Young's modulus, $E(\mathrm{MN/m^2})$	Density, $\rho(\mathrm{g/cm^3})$	Relative efficiency as a column: $\sqrt{E}/\rho$	Relative efficiency as a panel: $\sqrt[3]{E}/\rho$
Steel	210,000	7.8	59	7.7
Titanium	120,000	4.5	77	11.0
Aluminum	73,000	2.8	99	15.0
Magnesium	42,000	1.7	120	20.5
Brick	21,000	3.0	48	9.0
Concrete	15,000	2.5	49	10.0
Graphite-fiber composite	200,000	2.0	225	29.0
Bone (approx.)	18,000	2.0	67	21.1
Softwood (spruce)	10,000	0.35	290	61.4
Hardwood (oak)	12,000	0.65	170	35.5

about twice as efficient as steel, which is why aircraft wings and fuselages are covered with it. Magnesium is more efficient still, but this metal is too reactive chemically: it is liable to burn and to corrode.

An interesting consequence of the figures in this table is the way they govern the top speeds of aircraft. As aircraft speeds increase, the surface of the structure is prone to aerodynamic heating, largely because of the compression of the air around the leading edges of the structure. At supersonic speeds the temperature rise can seriously weaken the material of the airframe. The top speed of the Concorde, for instance, is chosen so that the temperature rise is compatible with the safe upper temperature limits of the aluminum alloys. Any substantial increase in speed would require the use of titanium, which has a much higher temperature resistance. But titanium also has considerably lower values of $\sqrt{E}/\rho$ and $\sqrt[3]{E}/\rho$. Thus the presence of titanium in any quantity in a projected hypersonic aircraft would result in a weight increase that would probably be unacceptable in a civilian machine.

At a lower level of technology, the table shows that brick, concrete, and other forms of masonry are reasonably efficient materials for constructing walls, which are primarily panels subjected to vertical compressive loads. Steel remains popular because it is fairly easy to expand the I value of the section; that is, steel can be made into tubes, channels, and H sections. When steel is wanted for panels it can be corrugated, or stringers can be riveted or welded on in such a way as to increase the resistance to buckling. In passenger automobiles the steel body panels are nearly always curved; curved panels offer greater resistance to buckling than flat panels do. In the manufacture of truck bodies and shipping containers, however, where the panels are usually flat, metals are being ousted by modern composite or wood-based materials that have much higher values of $\sqrt[3]{E}/\rho$, and so tend to produce much lighter structures.

In biology, bone is a better material for struts than steel, although bone is not as good as aluminum. But bone has other functions to perform than its purely structural ones. It is to a large extent self-healing, for instance, which cannot be said of aluminum. The marrow inside bones synthesizes red blood cells. Bones are depositories for unwanted calcium ions in the body—and so on.

Another really efficient material is wood. Although wood was not, of course, specifically designed for manufactured structures, the tonnage of timber used in the world today is considerably greater than that of all the metals put together. And as a biological material wood is even more impressive. From its high material efficiency criteria one can see why big trees are by far the largest of all living things. The values in the table also reveal why the largest and tallest trees are softwoods, such as the California redwood (*Sequoia gigantea*), and why softwoods of all sizes usually have more slender trunks than hardwoods.

For some time wood has been out of fashion with engineers—partly because of its genuine disadvantages, such as rotting and swelling, but partly because of the arrogance of "progress," which assumes that the newest thing is best. However, wood is coming back in a number of ways: in its natural state as traditional lumber, in various reconstituted forms bonded by modern adhesives, and as a model for the design of synthetic composites. As the table on page 51 shows, even the newest and most expensive of the synthetic fiber-resin composites is not really as efficient for making struts and panels as natural wood. Composites do, however, have other structural virtues that wood does not have; and they usually do not shrink, swell, or rot. Of this, more in Chapters 7 and 8.

TENSION

As we said in our discussion of weight-costs, when a light-to-moderate load has to be carried over a fairly long distance, the lightest and cheapest way of doing the job may well be to use an all-tension structure. Naturally, because action and reaction are equal and opposite, the tension forces have to be compensated by equal and opposite compressive ones. With a little low cunning, one can often arrange for this to be done for free by a preexisting substrate, by a fluid pressure, or by something else of that sort.

In a suspension bridge the tension in the main supporting cables is reacted by the ground under the bridge—a substrate that does not have to be constructed or paid for. In the same way spiders are able to spin light, cheap and elegant webs because the tensile loads are counterbalanced by the wall or other support to which the threads are attached.

Rigid external supports for tension devices are not always feasible, however, especially with mobile structures such as balloons or animals. In fact, the most common, and arguably most successful, of all structures is the living cell, in which a tension membrane encloses a core of fluid or jelly which reacts against the tension loads by hydrostatic compression.

As we shall see, the problem of evolving a material suitable for a cell membrane was a difficult one. Furthermore, because of the risk of bursting the membrane, the size of the largest cell that could exist with safety was not very great. The limited mobility of a unicellular animal—simple bladders, in effect—led to multicellular forms of life. Many soft, multicellular animals rely almost entirely on internal hydrostatic pressures to oppose the elaborate and varying tensions in their membranes. Some of these inverte-

Spider webs are light and efficient all-tension structures which depend on walls, twigs, and other preexisting supports to provide the necessary compressive reactions. If the spider had to make compression members as well as tension ones for its flytrap, its way of life would be quite impracticable.

Although there are strict limits to the size of unicellular animals, soft multicellular creatures can be very large. In the Pacific octopus (Octopus dofleini) tension loads and forces carried in (and sometimes generated by) the various soft tissues and membranes are opposed by hydrostatic compressive forces from within.

brate animals are very large; most of the bigger ones—like squids and octopuses—remain waterborne.

Although soft tension structures based on these principles have been immensely successful in nature, a long time ago the larger and more advanced animals acquired rigid compression members—bones, cartilage, and so forth—to counteract their tension forces. To minimize the weight added by these internal compression members, their number has been kept to a minimum.

For a great many years almost the only important manufactured all-tension structures were fishing nets—which, curiously, resemble spider webs. Latterly, technology has been returning to bladderlike tension structures, often with considerable advantage. As a matter of fact, the idea is not a new one in vehicular technology. Around 1000 b.c. the boatmen on the Tigris and the Euphrates were using inflated animal skins to construct rafts for the transportation of cargo. The idea seems to have taken a long time to catch on more widely, though, perhaps for the lack of a material other than leather for making membranes.

Blown-up "boats" made from rubberized fabric were sold to bathers between the two world wars; these were intended exclusively for light-hearted recreation. In an improved form they came into serious use in 1939 as rescue dinghies for aircraft. As such they have saved many lives. More recently, they have been developed by coast guards as large, power-driven lifeboats. These vessels have proven so efficient and durable that they are ousting traditional rigid lifeboats.

When I was involved in the development of aircraft rescue dinghies, I often heard the objection that such structures were dangerously vulnerable to punctures or tearing, but in my experience this seldom happened. In any case, failure by holing can be guarded against by the use of flexible internal diaphragms. Boats of this type are usually inflated with air or carbon dioxide. Around 1958 it occurred to two engineering professors at Cambridge University that fluids such as oil could be transported within large, flexible, waterborne bags resembling greatly enlarged wineskins or amoebas. When such containers were made, they worked very well. They were christened *Dracones* for their vaguely dragonlike appearance. They are used for shipping oil on the rivers of South America and elsewhere. (They are also used for carrying fresh water to the hotels on the Greek islands.)

A few years ago I happened to attend a meeting at the Ministry of Heavy Industry in Budapest. During a conversation afterwards I asked my hosts whether there was any interest in Dracones in Communist countries. They all assured me that there was no interest whatever in such eccentric capitalist ideas. A few minutes later I walked out of the ministry building

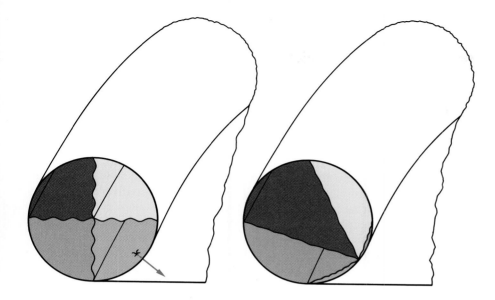

*Inflatable rafts, such as aircraft rescue din-
ghies, are not likely to sink if the side is
punctured. Flexible internal diaphragms divide
the wall into separate compartments. Should
the wall be punctured, the undamaged compartments inflate to compensate for the
collapsed one.*

and leaned over the parapet of a nearby bridge over the Danube. As I did
so a tug emerged from the arch below me, towing a Dracone barge, obviously full of oil. It is probably safe to assume that Dracones are used fairly
extensively on the rivers of Eastern European countries.

A new type of synthetic all-tension structure appeared toward the end
of the eighteenth century. The first successful manned airborne device, a
hot-air balloon, flew on October 15, 1783. This balloon was conceived,
developed and built in France by the brothers J. M. and E. J. Montgolfier.
Even though the balloon had already been tried out with animals, the
inventors apparently did not choose to make the inaugural flight themselves. The skin of this balloon was constructed of panels of linen *buttoned*
together. To make the thing more or less gas-tight, the envelope was lined
with paper. (The Montgolfier brothers were papermakers.)

The Montgolfier hot-air balloon was quickly followed—and for many
years superseded—by the hydrogen balloon, whose envelopes was generally made from rubberized fabric. (The first successful hydrogen balloon
was actually made from rubberized silk, which must have been expensive
in the 1780s.) Development was suprisingly rapid. The English Channel
was first crossed by air on January 7, 1785, in a hydrogen balloon manned
by a Frenchman, Jean Blanchard, and an American doctor from Boston,

*Blanchard and Jeffries crossing the Channel in
January, 1785.*

A zeppelen under construction, around 1926.

John Jeffries. They took off from Dover with a fair wind, but before they reached France the balloon developed a gas leak. To avoid ending up in the sea the aeronauts jettisoned not only all the ballast and loose gear in the balloon but most of their own clothing, including their trousers, as well. They descended near Calais, scantily clad, late that winter afternoon.

These early soft balloons were very effective as airborne vehicles, but they were wholly dependent on the wind. Attempts to make soft *dirigible* (steerable) powered balloons ("blimps") which would be navigable against the wind were only moderately successful. Blimps were used extensively for convoy escorts and antisubmarine patrols during World War I; however, they were largely abandoned after 1918 as being too slow and vulnerable. Rigid zeppelin airships, whose hulls were fitted with elaborate systems of girders capable of taking compression, have fallen prey to all the difficulties and complications that inevitably beset compression structures with very low structure loading coefficients. If large commercial airships are ever to be a practical proposition, the whole problem of their structure will have to be reconsidered in some more inventive way.

The hydrogen balloon did play an important part in World War II, when large numbers of stationary or tethered balloons were used to protect European cities from attack by low-flying bombers. The Montgolfier balloon has experienced a major revival in recent years for recreational flying, because hot air is cheaper than helium and safer than hydrogen. (When the wind is in the right direction Montgolfier balloons pass over the author's garden in England in suprising numbers, sometimes at altitudes and speeds low enough to permit lengthy conversations with the aeronauts. For those on the ground, at least, balloons are blessedly silent!)

The idea that a balloon could be used not only as a flying machine but also as a building is credited to the distinguished and original British engineer F. W. Lanchester (1868–1946). Lanchester was primarily an aerodynamicist. As a result of his calculations and experiments, dating from about 1894, he deduced the *circulatory theory of lift* to explain why birds and airplanes are able to fly. His paper on the subject was refused in 1898 by the reputable scientific journals, and although he published his ideas in book form in 1907, they were ignored or scorned by scientists, engineers, and aircraft pioneers. Under the pressures of World War I, Lanchester's ideas were rediscovered by Ludwig Prandtl in Germany in 1915. The Lanchester-Prandtl circulatory theory is now the basis of all modern aerodynamic calculations.

Another of Lanchester's contributions was the idea that an *air-supported roof*, a contrivance resembling a balloon pegged to the ground, might be used as temporary building to house exhibitions. He put forth this idea in a paper (which he did manage to get published) in 1910, though again no-

body paid any attention. The patent, which he took out in 1918, quietly expired.

Practical exploitation of the idea did not take place until the 1950s when the American firm Birdair developed air-supported housings for the radar screens of the Dewline, the distant early warning system that stretches across the North American Arctic. Birdair went on to develop the concept for warehouses, sports stadiums, and so on. At least 50,000 such buildings, with spans up to 700 feet, have now been erected throughout the world. These "bubbles" are not entirely ephemeral; they have an anticipated life of around 50 years. The use of air-supported roofs has drastically reduced the cost of warehousing in the United States.

Air-supported buildings are kept inflated by a fan that runs more or less continuously, and alternative blowing arrangements are needed in case of failure of the main fan. These buildings have to be entered and left by means of an air lock, an intermediate chamber between the outside air and the pressurized air within. Furthermore, in large bubbles especially, the material of the roof membrane needs careful consideration. Even with all these requirements, air-supported roofs are much cheaper than conventional roof structures.

The single most important artificial all-tension structure is probably the pneumatic tire. Like other important inventions, its development was delayed for many years. The pneumatic tire was invented, tested, and patented by a 23-year-old Englishman, R. W. Thomson, in 1845. Thomson's tire was a remarkable technical success, but 1845 was the period of the railway boom. The powerful railroad interests combined with the powerful horse interests to promote reactionary legislation that delayed the development of mechanically propelled road vehicles for 50 years.

During Thomson's lifetime, there was little demand for pneumatic tires for horse-drawn vehicles. In 1888, when Thomson was dead and his patent had expired, the idea was revived by J. B. Dunlop for use on his "pedal cycles." Pneumatic tires made the bicycle an enormous success, and Dunlop made a large fortune. Modern fast road transport is possible only because of Thomson's invention. Pneumatic tires have also made land-based aircraft practicable; without their shock-absorbing capabilities, today's airplanes might have all been seaplanes.

TENTS AND SAILING SHIPS

MIXED TENSION AND COMPRESSION STRUCTURES

Although all-tension structures have been very successful, especially in biology, for many applications they are too flexible or otherwise unsuitable. It is often expedient to counteract the tension forces within a structure by inserting solid struts rather than by relying on a fluid pressure. The number of struts should generally be kept to a minimum, as we indicated in our discussion, "The Effects of Subdividing a Structure," on pages 48 and 49.

The rigs of sailing vessels provide an interesting example. Sails have played an important role in history and they may do so again as energy gets more and more expensive. It is difficult to see how sails could stand up against the winds of Heaven supported by any structure drastically different than the traditional one, in which compressive loads are collected into one or more struts or masts supported by a plurality of ropes or tension members such as shrouds and stays. The fewer the masts, the greater the structural and aerodynamic efficiency. Thus a small yacht or dinghy is usually rigged with a single mast, as a

sloop or a cutter. For small vessels this is a better arrangement than the multimasted design of a ketch or a schooner.

As ships got bigger and more masts were added, reduced labor costs sometimes compensated for the increased weight and cost of the rig and the loss of aerodynamic efficiency. It is appropriate that the schooner rig was invented in America, where labor has always been more expensive than in Europe. The first schooner is supposed to have been built in Gloucester, Massachusetts, in 1713. Previously the European solution to the design problem presented by the larger sailing ships was some form of square rig. Although a square rig has considerable advantages—less chafe, bigger sail area in relation to spars, more easily managed by a skilled crew—it is labor-intensive. Square-riggers require larger crews than schooners.

The original schooners had two masts, as do most small schooners today. As ships got bigger it was necessary to subdivide the rig still further so as to make it manageable by a small crew. During the nineteenth century, New England produced three-, four-, five-, and six-masted merchant schooners in considerable numbers.

This trend towards subdividing and subsubdividing the rig culminated in 1902 with the world's one and only seven-masted schooner,

The Thomas W. Lawson.

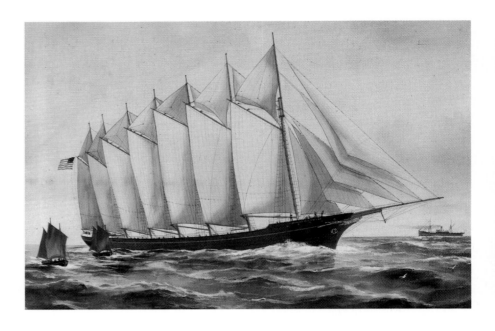

The Schlumberger Research Center in Cambridge, England, is largely a tension structure.

the *Thomas W. Lawson.* This vessel was more than 400 feet long and each of her seven masts was 193 feet tall. All the gear for the masts was led to two steam winches, an arrangement that enabled the ship to sail with a crew of only 16 hands. Although she was no doubt cheap to run, subdivision—with the corresponding loss of structural and aerodynamic efficiency—had probably been taken too far, and the Thomas W. Lawson does not seem to have been a very competent sailing ship. She was lost on the Scilly Islands off the coast of England in 1907.

Like sailing ships, tents are a very old idea. The English word comes from the Latin *tenta,* which in turn derives from the verb *tendere,* meaning *to stretch.* In other words, the tent has always been a tension structure. Because it is designed to be portable, the compression loads are concentrated into one or a few poles and the tensions diffused into various membranes and guy ropes.

As a habitation, the ordinary tent has the disadvantage that the canvas walls are too flexible and tend to flap in the breeze. In recent years, various designers, such as Professor Edmond Happold of the University of Bath, England, have taken the tent concept in hand and enlarged and modernized it. The poles remain poles, or large struts, but the canvas walls have been replaced by semirigid tension panels made of plywood or plastic which does not flap in the wind.

Lately these modern tent buildings have been erected almost all over the world in considerable numbers. Many of them are so large that they are not portable in the ordinary sense. Unlike some innovative structures, these rigid tents can be very beautiful. They are rather like Gothic cathedrals turned inside out, with all the arches and vaultings put into tension rather than compression.

TRUSSES AND BEAMS

We turn now from air-supported tension devices to some of the heavier hardware of tension and compression structures: trusses and beams. The word *truss* derives from the French *trousser*, meaning to *tie up*. In technology, a truss is a tied, braced, or latticed structure. The word *beam* comes from the German *Baum*, for tree, and its identical meaning in Old English survives today in the tree means names *hornbeam* and *whitebeam*. (The nautical term *boom* is another version of *Baum*.) Nowadays a beam is defined as a horizontal support within a structure. In more primitive technologies, the most common beams have been those made from whole tree trunks. We still sometimes refer to a ridgepole—the topmost beam of a roof or the topmost horizontal pole of a tent—as a *rooftree*

A beam is, in a sense, a solid truss: the beam generally does much the same job as a truss, although the beam has a continuous cross section and is not strutted or braced. In one guise or another beams are very common indeed in nature, and in technology they are certainly the most ubiquitous of all structural forms. However, as we have seen, Galileo, rather excusably, got his beam sums wrong. Many eminent scientists and mathematicians who followed him either got it wrong again or else produced only partial solutions.

Modern beam theory took a long time to appear in its present straightforward form, and it evolved from the work of many different people. This was partly because a generalized analysis of the strength of beams depended not only on acceptance of Newtonian mechanics, but also on familiarity with the language of stress and strain, which, as we have seen, was not in common use until the 1830s.

In a beam made from a material that obeys Hooke's law, the distribution of stress varies linearly across the thickness of the beam. On one side

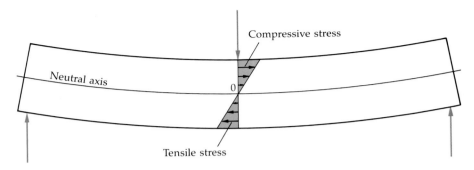

Distribution of stress through the thickness of a beam.

Shapes of various beams used in engineering to maximize resistance to bending at their surfaces; from left, an I-beam, a Z-beam, and a rail.

of the beam the stresses are tensile; on the opposite they are compressive. The transition from tension to compression through the thickness of the beam proceeds in a linear fashion. This is the case regardless of the shape of the cross section of the beam (e.g., round, square, tubular, I section, or asymmetrical), always provided that the material behaves in a Hookean way.

Somewhere inside the beam, usually near the middle, there is a line (actually a plane if we are thinking in three dimensions) along which the longitudinal stress in the beam is zero. This is known as the *neutral axis*. The neutral axis always passes through the centroid—sometimes called the center of gravity—of the cross section of a beam. At a point that is a distance y from the neutral axis, the longitudinal stress σ (sigma) in the material of the beam is My/I, where M is the bending moment to which the beam is subjected at that point, and I is the second moment of area of the cross section. Clearly the stress is greatest where y is greatest—as far as possible from the neutral axis: in other words, at the outer surfaces, the top and bottom, of a beam. Thus, when a beam is bent to destruction, it will usually begin to fail at an outside surface and the fracture will spread inward. Likewise, the material farthest from the neutral axis of an intact beam is most crucial in resisting bending; material close to the neutral axis will usually be taking little stress and, quite literally, not pulling its weight.

For a beam to resist bending, therefore, it should have some kind of expanded cross section; that is, one with a high I value, or second moment of area. If the load, and so the bending moment, is more or less in one direction, it behooves the engineer to use a beam shaped in an I section or similar configuration. If, however, the load may come from any direction, it is generally better to use a beam shaped like a round tube. This is why straw, bamboo, and many bones have a round, hollow section. (We shall see why trees, in apparent contradiction of this principle, are "solid" in Chapter 7.)

There is often a close resemblance between the shapes of beams, which have to resist direct *bending loads*—that is, loads applied to their sides—and long columns, which have to resist compression along their lengths. Columns are liable to fail by buckling, which is a form of bending. Trees and bamboos have to resist vertical compression loads arising from their own weight and that of any snow, ice, or other material they may have to support, and also bending loads caused chiefly by the horizontal pressure of the wind. The two requirements are sufficiently similar from the structural point of view to enable these plants to invest their metabolic energy efficiently and thus grow tall.

Beams are often classified according to where the load is applied and whether their ends are free or fixed. For example, a *cantilever* is a beam or truss that has one end fixed while the other end sticks out to support a load. In biology, all trees and most plants are cantilevers, as are horns, tusks, claws, arms, legs, and fingers. In technology, monoplane wings and most other features that project from a firm base are also cantilevers.

A *simply supported beam* is a beam or a truss that is propped, but not clamped, at both ends, allowing it to support a load in the middle. The backbones of quadruped animals are, in effect, simply supported beams. Examples from construction include floor joists in houses and many bridge spans. A simply supported beam may be regarded as two cantilevers put back to back and upside down, with the support becoming the load and the loads becoming supports.

Other beams are not supported "simply." They may be supported at any points along their length, as is generally the case with ordinary house floorboards, which rest on a dozen or so joists. The situation is further complicated if a beam is fixed or clamped (*encastré*) at one or both ends or at any intermediate supporting point.

Trusses generally have more elaborate structures than beams. In one form or another, trusses have been used in shipbuilding from very early times. We have already mentioned mast stays, the tension members used to counterbalance the vessel's main compression member(s). Trusses were also used on the hulls of boats: when boatbuilders moved away from com-

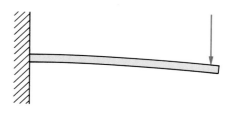
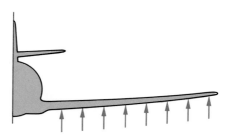

Cantilevers. At top, a cantilever with a concentrated load. Below, a cantilever—an aircraft wing—with a distributed load.

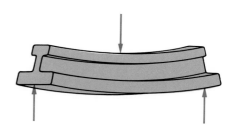

A simply supported beam.

One cantilever

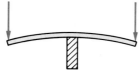

Two cantilevers

Two cantilevers upside down

A Mississippi riverboat, the John A. Wood, around 1905. The hogging trusses are the thin cables visible arching across the top of the boat. The verb to hog means to arch the back upward the way a hog does. Applied to a ship, to hog means to arch the ship upward in the center as the result of a strain.

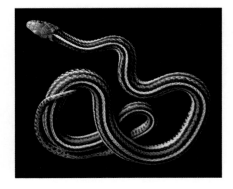

The Fink-truss arrangement of ribs, tendons, and muscles enables the snake to move about by changing its shape.

paratively strong and rigid dugout or hollowed tree trunks, their fabricated hulls were weak, flexible, and leaky. Even on the smooth waters of the Nile, the hulls of ancient Egyptian ships needed *hogging trusses*. The hogging truss was revived during the nineteenth century for American river steamers, which were lightly built and hard driven, and so had a similar problem.

The first important technological use ashore for large trusses arose with the need for railroad bridges. Although a suspension bridge is usually the lightest and cheapest bridge for large spans, it is too flexible when subjected to the loads imposed by trains and locomotives. Some kind of a beam bridge was needed. One of the earliest types of beam bridges was the Fink truss, in which crisscrossed diagonal tension members supported vertical compression members.

As a matter of fact, the Fink truss had already been exploited by nature for many millions of years in the bodies of vertebrate animals. By making use of contractible muscles as tension members, vertebrate animals (especially snakes) are able to transport their weight from one location to another by changing their shape, sometimes in a rather spectacular manner. At the other end of the vertebrate spectrum, a horse may be regarded as a rather successful bridge: its large body, capable of bearing a substantial load in addition to its own weight, rests on four slender compression members supported efficiently by an assortment of tension members (tendons, muscles, skin).

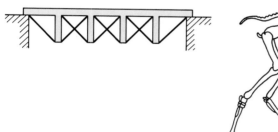

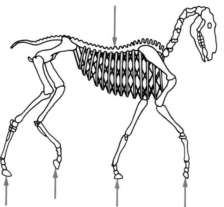

Left, a Fink-truss bridge near Lynchburg, Virginia. Its span is 52½ feet. The vertical and diagonal members are of wrought iron and the supports for the floor system are of 15-inch-thick wood. Below left, the basic configuration of a Fink truss. Below right, the musculoskeletal systems of many vertebrate animals resemble Fink trusses. In the horse, the vertebrae and ribs form the compression members; the space between the ribs is crisscrossed by a network of muscles and tendons.

But railroad bridges and similar technological structures should not wriggle like snakes or dachshunds. A Fink truss becomes a more rigid structure when it is lined along the bottom with a series of tension members. The result is a Pratt or Howe truss. Such trusses, stayed perhaps with wires, are well suited for structures with low structure loading coefficients— that is, for structures designed to handle comparatively light loads over comparatively long spans. These features were typical of the wings of the early airplanes, and the Pratt truss was employed for them: hence the biplane design. The strutted biplane was used by the Wright brothers and continued to be built well into the 1930s. A few are still to be seen about today.

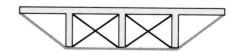

A Pratt or Howe truss is a modification of the Fink truss. Additional stability is achieved by installing a continuous tension support along the bottom of the truss.

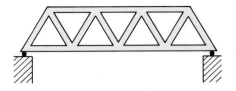

Above, a Warren girder is a zigzagged simpli-
fication of a Pratt truss. Right, modified
Warren girders support the wings of the U.S.
Navy Stearman biplane.

For some engineering purposes, the Pratt truss is modified to the zig-zag form known as the Warren girder. This construction, which is suited to fabrication from rolled steel joists or other metal members, is widely used for modern bridges. Pratt trusses and Warren girders are rare in biology, although occasional examples do occur in animal skeletons.

MONOCOQUE OR SHELL STRUCTURES

Sweet Echo, sweetest nymph, that liv'st
unseen
 Within thy airy shell . . .

JOHN MILTON
Comus

Milton could scarcely have foreseen the existence of modern aircraft or flight attendants, but there are an enormous number of "airy shells" flying around today. Practically all modern aircraft are shell structures, and so are ships and most automobiles. Nowadays many of us are in the habit of thinking that only shell structures—preferably made from plastics or light, shiny metals—are truly modern and efficient. More traditional structures, braced or latticed ones, are considered old-fashioned and fuddy-duddy.

In fact, shell structures are not especially successful in the larger and more sophisticated forms of life. Among plants, grasses and bamboos might be deemed shell structures, and so might many flowers and seed-pods. But trees, which are the largest and longest-living of plants, are not shell structures. Among animals, many insects have hard shells, as do lobsters and shellfish. Of the more advanced animals, turtles, armadillos, and tortoises have thick shells, but these serve as defensive armor; the animals have an ordinary skeleton inside.

The functions of a closed rigid shell can be summarized as follows:

1. *To provide protective armor.*
   In some cases, a barred cage structure (e.g., a ribcage) does better.

2. *To contain or to exclude liquids or gases.*
   This is the function of ships' hulls and the pressurized fuselages of aircraft and spacecraft. Nature, however, generally prefers to use a flexible bag. Fuel tanks in aircraft are often bag-tanks modeled on the biological strategy.

3. *To carry axial compressive loads.*
   The leg bones of vertebrates meet this need, as do many engineering structures, such as the legs of oil rigs. (We have already described these features as tubular beams, but they could just as well qualify as shell structures.) Often, however, stayed struts are lighter and cheaper.

4. *To provide a smooth surface in order to minimize aerodynamics or hydrodynamic resistance.*
   This is an important consideration in ships and aircraft, but as a way of ensuring a smooth surface, the rigid shell is rare in animals. Birds prefer feathers; fish and whales make use of a cushion of skin stretched over soft flesh. These flexible arrangements make it possible for minor dents and roughnesses to smooth themselves out automatically.

5. *To enable the structure as a whole to resist shear and torsion loads.*
   From the point of view of the engineer, this is often the heart of the matter, as we shall see in the remainder of this section.

In aircraft, twisting of the wings, fuselage, or control surfaces (ailerons, elevators, rudder) can have disastrous results. The need to provide adequate strength and stiffness against shear and torsion (a wind-generated twisting force) is usually the major weight-producing factor in the design of modern aircraft.

In both plants and animals, nature is astonishingly clever at evading the need to resist shearing or twisting forces. The reason why no animals, even soaring birds, exhibit any significant requirement for strength and stiffness against shear and torsion is still something of a mystery. To the best of my knowledge, no one has undertaken a serious investigation of the question. When we consider the difficulties that shear and torsion have presented to engineers, it seems likely that the solution to these problems must have been of great importance in the evolution of animal morphology.

Shell structures are common in smaller forms in nature, as in the lobster.

In traditional technology, not only craftsmen but designers turned a blind eye to the destructive potential of these forces. Most antique furniture wobbles and comes loose at the joints, but these flaws have generally not been life-threatening. On the other hand, the fact that virtually all wooden ships constructed before 1830 leaked was a serious matter indeed. In the ships of England's Royal Navy, leakage due to shear distortions was finally reduced when Sir Robert Seppings introduced diagonal iron bracing into wooden hulls around 1830. However, when I was working as an apprentice in a wooden shipyard on the Clyde, just over a hundred years later, it was very obvious that the shipwrights, however skilled they may have been with an adze, had no appreciation whatever of the effect of shearing forces on a wooden hull. In fact, things had changed very little since the time of St. Paul.

Although shear stresses have been generally ignored by most ordinary people from the earliest times to the present day, to the modern engineer they constitute a very serious problem. To the aircraft designer they can be a nightmare. Sophisticated structures like aircraft need to have not only adequate shear strength, but also sufficient stiffness in shear and torsion. Otherwise, the wings or tail surfaces may be twisted off under the aerodynamic forces to which they may be subjected.

The act of twisting a hollow boxlike or tubular structure such as a fuselage or a wing must set up a shearing force in the skin. This shearing force can be resisted either by inserting a lattice inside the structure or by surrounding the structure with a continuous skin or *monocoque*. From the point of view of strength, there is much to be said for the lattice. In the old-fashioned biplanes, the double wing with its struts and wire braces provided a torsion box—that is, a frame strong and stiff enough to resist the various twisting forces that arise during flight. However, for a given stiffness or rigidity in shear, the lattice structure is likely to be much heavier than one covered with a continuous metal skin. That is why virtually all modern monoplanes, with their single thin wings and their much higher speeds, are covered with shiny aluminum plates.

Stiffness in torsion is governed by the shear modulus G, which is the shear analogue of the Young's modulus E in tension. As we mentioned in Chapter 2, for a continuous material such as a metal panel in a wing or a fuselage, G is related to E as follows:

$$G = \frac{E}{2 (1 + \nu)}$$

where ν is Poisson's ratio. Because for most metals ν is approximately 0.3, the formulation becomes

$$G \approx \frac{E}{2.6}$$

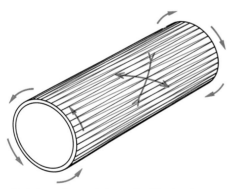

Twisting a long tubular or boxlike structure sets up a shearing force in the skin.

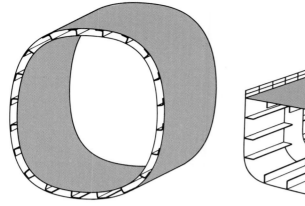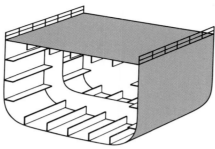

In aircraft fuselages (left) both stringers and ribs are added to the skin to stabilize the fuselage against buckling. In a normal steel ship (right) the skin plating is stabilized by decks, stringers, ribs, and the keelson.

A shear modulus can be calculated for a lattice structure as well. Even in the most favorable circumstances, G for a lattice cannot exceed $E/4$. Thus, for a given torsional stiffness, a lattice structure may be nearly twice as heavy as the equivalent monocoque. This is the chief reason why modern aircraft wings and fuselages look the way they do.

The loads upon sophisticated shell structures such as aircraft are complex and involve tension and compression as well as shear. Because thin

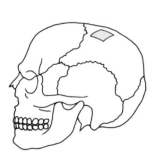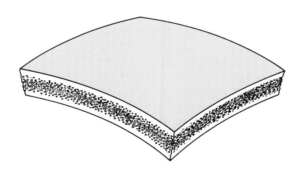

Cancellous bone in the human skull offers an example of sandwich construction in nature.

panels are prone to Euler-type buckling under both compression and shear, the panels of monocoque structures nearly always have to be stabilized. It is generally done by means of a mesh of internal ribs and stringers (longitudinal stiffening members), a system that is widely used and generally satisfactory. A similar system of ribs or stringers is used in nature to stabilize the walls of plants such as grasses and bamboos.

The only widely used rigid monocoque found in vertebrates is the skull, which is a sort of armored box to contain the brain. The walls of this structure are stabilized against buckling, but not by a system of ribs and stringers. The skull uses a *sandwich* construction, in which the main shell wall is split into two layers separated by a low-density core. The core takes the form of *cancellous* bone—that is, bone containing many bubblelike voids.

THE CONVERGENCE OF STRUCTURAL FORMS

It is a recurrent curiosity of the study of structure that apparently unrelated natural and artificial devices sometimes bear remarkably close resemblances. The first scientist to explore seriously the relationship between biological and technological structures was Sir D'Arcy Thompson, a Scottish biology professor. In a 1917 book called *On Growth and Form*, which became a classic in its day and still is, Thompson advanced the idea that animals resemble engineering structures and obey the same laws of mechanics, drawing parallels between vertebrates and bridges. Partly because his knowledge of engineering theory was rather sketchy, and partly because he was way ahead of his time, his book was regarded as an amusing curiosity by many biologists.

Another interesting and amusing case arises from the parallel between vertebrate animals, such as humans, and square-rigged ships, such as clippers. The compression loads are concentrated into vertebrae and ribs in the former, and into jointed masts and yards in the latter. Tension loads are diffused into tendons, skin, and other membranes in humans, and into ropes and sails in the ships. The humans have a small monocoque, the skull, at the top of the structure; the ship has a larger and more vulnerable shell, the hull, at the bottom.

Such similarities can be quite instructive. In fact, the whole subject of the convergence of structural forms in nature and in technology would bear detailed examination in the light of modern knowledge. Engineers and biologists might learn a great deal from one another in the course of making comparisons.

The first important use of a sandwich construction in engineering was in the Mosquito aircraft, which were famous in World War II. The "Mozzy" was, of course, an all-wooden plane. The inner and outer skins of the wings and fuselage were made from ordinary birch plywood. These skins were usually separated by a core of balsa wood, a timber of very low density. The Mosquitos were immensely successful in various operational roles, and 7,781 of them were built. A few are still to be seen in museums; there is one in the National Aviation Museum in Ottawa, for instance. Nowadays, balsa wood has been replaced as a core material in sandwich constructions by various light-weight synthetic materials such as plastic foams and honeycombs.

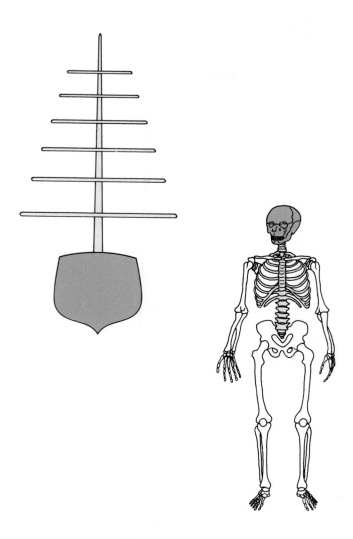

The convergence of structural form: like causes tend to produce like results. The resemblance between the structure of the traditional square-rigged sailing ship and the structure of the human body is startling. (Tension members are omitted.)

Although rigid monocoques are currently fashionable in engineering and one can see them in considerable numbers at every airport and in every harbor, they represent quite a recent innovation in technology. Their future is doubtful, not because they are inefficient as structures, but because they may turn out to be too dangerous due to the ease with which cracks can spread across them. We will examine this drawback of shell structures more closely when we discuss the modern science of fracture mechanics in Chapter 4.

4

TENSILE STRENGTH AND TENSILE FAILURE

Cohesion and Modern Fracture Mechanics

The race is not to the swift, nor the battle to the strong.
A living dog is better than a dead lion.

ECCLESIASTES

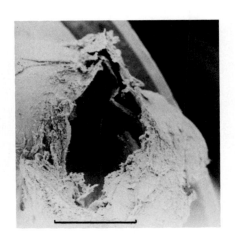

Tensile failure is not just an engineering phenomenon. Though the structural integrity of the arteries enables them to withstand 100,000 heartbeats a day, the artery walls may form a bulge called an aneurysm, which may burst, like this cerebral aneurysm. The scale bar represents 1.0 mm.

Although an increase in the tensile strength of materials would not reduce the weight and cost of structures nearly as dramatically as people sometimes imagine, tensile strength is a very important property. Tensile failure—the breakage of structures under tension—is the cause of a high proportion of structural accidents: in one way or another, it has been responsible for an enormous number of deaths, historically and in modern times as well. This is true not only in technology but also in medicine. About one percent of human deaths in developed countries today are due to the bursting of blood vessels in the brain. That is what killed the writer Robert Louis Stevenson at the age of 44. Tensile failures in animals and plants must be fairly common.

Until recently, however, the subject of cohesion and fracture was neglected or avoided by academic scientists. Engineers, of course, have always been compelled to take seriously the question of the practical strength of their materials. They have done an enormous number of tensile tests, mostly on metals, and reduced the results to statistics such as "allowable stresses" for tables in handbooks. But the engineers' approach was entirely a pragmatic one, and no generalized theory of the strength and weakness of solids existed until after World War I.

The breakthrough, owing to a marriage between science and engineering, was the work of a young Englishman named A. A. Griffith (1893–1963). Griffith began his work on the strength of solids around 1918 at the Royal Aircraft Establishment at Farnborough, England. (Under a veneer of officialdom, this government laboratory has often shown a praiseworthy tolerance towards eccentric and original young people.) His assistant was a slightly older man called Lockspeiser. For various reasons, I never knew Griffith himself, but many years later, Lockspeiser—by that time Sir Benjamin and the retired Chief Scientist of his ministry—told me quite a lot about the background of Griffith's work at Farnborough.

It would be interesting to have a well-researched "history of the acceptance of scientific ideas." Resistance to the theories of Copernicus and Darwin was, of course, mainly theological. But this would not account for the neglect of Mendel's work on heredity nor for that of Lanchester's transformation of aerodynamic theory. Still less does it account for the long delay in the acceptance of Griffith's ideas about the fracture of solids. I am afraid that both scientists and engineers are sometimes intellectually conservative and unwilling drastically to revise their basic concepts.

Griffith's seminal—indeed, revolutionary—paper on strength and fracture was published by the Royal Society in 1920. It was never refuted, or even strongly objected to. It was simply neglected, for at least a generation, by the people who should have studied it avidly. As it happened, the author worked at Farnborough throughout World War II in the same building in which Griffith had done his research, and I must have used the same testing machines. Although we were concerned—desperately concerned—with the strength of materials and with the failure of aircraft, I seldom heard Griffith's name mentioned; certainly we made little or no use of his ideas. We were vaguely aware that a man called Griffith (who was then working in the engine department) had done some experiments on the strength of glass, but we had no conception at all of the revolutionary nature of his ideas, which were applicable not only to the breakage of small specimens of materials like glass rods but also to the failure of large structures such as airplanes and ships. The recognition of the importance of modern fracture mechanics—the field that Griffith pioneered—began to dawn only around 1954, after the pressurized fuselages of three Comet airliners had exploded in the air with considerable loss of life.

In defense of the scientific establishment, it can be argued that Griffith's now famous paper is curiously difficult to understand. In style it is reminiscent of the writings of Thomas Young a century earlier. Unless one reads Griffith's paper very carefully, it is easy to miss his points altogether. In the following sections I have simplified Griffith's arguments to some extent for the sake of clarity.

GRIFFITH: THE THEORETICAL STRENGTH OF SOLIDS

It is obvious from looking at bubbles and droplets that liquids have a surface tension: they resist having their surface areas increased. If there were no surface tension, neither bubbles nor droplets would form. In order to increase the surface area of a liquid, the total number of atoms or molecules

The formation of bubbles depends on the surface tension of a liquid.

present at that surface must be increased. These additional atoms or molecules have to come from the interior of the liquid, and have to be dragged from that interior to the surface against the resistance of the various interatomic and intermolecular forces that tend to keep them there. The transference of the atoms or molecules requires energy, so a force must be exerted to extend the surface of the liquid—to, say, enlarge a bubble or flatten a droplet. Hence a liquid has both a surface tension and a surface energy. When a liquid freezes and becomes a solid, this solid is usually too rigid to be deformed—to be pulled into droplets or bubbles—by the surface tension. But the same surface tension—and more importantly, the same surface energy—is still there, even in the hardest solids. In fact, the surface energy of many solids is considerably higher than that of most common liquids.

When a solid is broken by the application of tension, at least two new surfaces have to be produced. Thus the minimum energy required per unit of cross-sectional area to cause the simplest tensile fracture must be at least twice the surface energy of the material. In order to estimate the theoretical tensile strength of solids—that is, the maximum stress they could withstand—Griffith equated twice the surface energy of the cross section to the strain energy (see Chapter 2) needed to separate two adjacent layers of atoms of the material. The strain energy η per unit volume in a stressed material is

$$\eta = \tfrac{1}{2}\epsilon\sigma$$

where η is the energy in joules per cubic meter, ϵ is the strain, and σ is the stress in newtons per square meter. For a material that obeys Hooke's law, this equation becomes

$$\eta = \frac{\sigma^2}{2E}$$

where E is Young's modulus. The strain energy per square meter between two adjacent layers of atoms that are initially x meters apart is thus

$$\eta = \frac{\sigma^2 x}{2E}$$

Equating the strain energy η to twice the surface energy γ, we get

$$\sigma_{\max} = 2\sqrt{\frac{\gamma E}{x}}$$

where $\sigma_{\max}$ represents the maximum stress the material can withstand without breaking. The assumption is that the interatomic bonds obey

Hooke's law all the way up to total failure. This is not really the case; in fact, the interatomic force curve bends in a way that is roughly parabolic, so that it is approximately true to predict for the theoretical strength of most solids that

$$\sigma_{max} = \sqrt{\frac{\gamma E}{x}}$$

For his classical experiments, made around 1919, Griffith decided to use ordinary glass. This was partly because glass is brittle and its fracture is generally less complicated than that of most metals. More importantly, it was easier to determine the surface energy with glass. Crystalline solids such as metals rearrange their molecular structure drastically when they harden, so that surface tensions measured on a liquid metal are not necessarily convertible into the surface energy of the solid state. Glass, however, is not normally crystalline, and when it hardens from the liquid or soft state there is no significant rearrangement of its molecular structure. It is reasonable, therefore, to suppose that surface tensions measured on hot, soft glass can be extrapolated to values at room temperature without any important error.

Griffith and his assistant Benjamin Lockspeiser measured the surface tension of their glass at various temperatures between 1110°C and 745°C. They extrapolated their measured values to about 0.56 newtons per meter at 15°C. Because surface tension in newtons per meter is equal to surface energy in joules per square meter, their result was equivalent to surface energy of 0.56 joules per square meter—which is in line with modern values. The Young's modulus of their glass was found by a simple mechanical test to be about 62,000 meganewtons per square meter (MN/m^2) or 9.0×10^6 pounds per square inch (p.s.i.)—a figure only slightly on the low side for glass, which usually has an E value around 70,000 MN/m^2.

If we assume the interatomic spacing in glass to be about 2.0 angstroms (1 angstrom = 10^{-10} meters), then the theoretical strength of Griffith's glass from the formula we have just derived is 1.3×10^4 MN/m^2 or 1.9×10^6 p.s.i. The formulas Griffith actually used were more complicated but more vague; with them, Griffith predicted a value of between 1 million and 3 million p.s.i. However, when millimeter-thick rods of the glass were actually tested in tension, they broke at an average stress of 170 MN/m^2 or 24,900 p.s.i.—a value nearly 100 times less than the theory indicated.

Griffith then drew his glass rods down into thinner and thinner fibers which, after cooling, he broke again in tension. As the fibers got thinner, their strength increased. The increase in breaking stress was fairly slow at first, but when the fibers became really thin their strength grew very rapidly indeed. Griffith was unable to test fibers thinner than about 10^{-4} inch;

even if he had been able to make such fibers it would have been difficult at the time to measure their thickness with any accuracy. Griffith's thinnest fibers showed a strength of almost 3,400 MN/m^2 or 500,000 p.s.i.—a twentyfold increase over the strength of the bulk glass. Moreover, when he came to extrapolate his experimental results, he found that the strength of fibers of vanishingly small diameter should be around 1.1×10^4 MN/m^2 or 1.6×10^6 p.s.i.—a value startlingly close to the theoretical one.

Since Griffith's time glass rods and fibers made from a material with a slightly higher Young's modulus have exhibited tensile strengths of rather more than 13,600 MN/m^2 (2.0×10^6 p.s.i.), a figure that agrees closely with the theoretical value.

These high strengths are not confined to glass. In 1952, two American researchers, C. Herring and J. K. Galt, showed that thin "whisker" crystals of a metal—tin, which is usually soft and weak—could be very strong. A couple of years later, working at Farnborough, the author found that crystals of various nonmetallic substances, such as hydroquinone, also showed approximations to the theoretical strength. The theoretical strengths of a number of miscellaneous solids are given in the table below. Although these theoretical stresses vary a good deal, they are all high. Even sodium chloride—common salt—has a theoretical strength that is many times greater than the practical strength of high tensile steel. These high theoretical values are not just a scientific abstraction; they have been reached or approximated experimentally many times, in many different laboratories. Why, then, is the effective strength of bulk materials in practice—that is, their strength in real-world contexts—almost invariably so much lower

Theoretical tensile strengths of various solids

Material	Surface energy (J/m^2)	Young's modulus (MN/m^2)	Theoretical tensile strength (MN/m$^2 \times 10^4$)	(p.s.i. $\times 10^6$)
Iron	2.0	210,000	4.6	6.7
Copper	1.65	120,000	3.1	4.5
Zinc	0.75	90,000	1.8	2.6
Aluminum	0.90	73,000	1.8	2.6
Tungsten	3.0	360,000	7.3	10.6
Diamond	5.4	1,200,000	18.0	26.3
Sodium chloride	0.115	43,000	0.62	0.9
Aluminium oxide	4.6	420,000	6.7	10.0
Ordinary glass	0.54	70,000	1.4	2.0

than their true cohesive strength—that is, the total strength of their inter-olecular bonds across the weakest section? As Griffith said in 1920, a single chain of molecules must have the theoretical strength, or else no strength at all. The problem was to explain the weakness of larger specimens.

Griffith's theories and explanations really fall into two categories: force or stress arguments and energy arguments. Intelligent engineers understood and accepted the force arguments comparatively early. The energy arguments, which are more important, more fundamental, and more widely applicable, met with an emotional resistance and were ignored or misunderstood for quite a long time. Even today they are by no means universally accepted.

We shall examine the force or stress arguments first.

THE FORCE ARGUMENT: STRESS CONCENTRATIONS

As the nineteenth century progressed, it became increasingly customary for engineers and mathematicians to attempt to calculate the distribution and magnitude of the stresses in structures like machinery, bridges, and ships. These calculations were made in a very broad and generalized way. In doing their sums, both the engineers and the mathematicians showed a rather lordly disregard for the actual geometrical details of the structures. A ship, for instance, would generally be regarded as a more or less uniform shell or tube: discontinuities like hatchways, portholes, and so on were ignored. Engineers provided for any weakening effect that these openings might have by incorporating a "factor of safety" into the calculations.

Use of a factor of safety meant that the highest allowable calculated stress in a structure was many times less than the known ultimate practical breaking stress that had been measured on samples of the material. In a ship, this factor of safety was generally 5 or 6; in machinery, it was often higher; and in parts of locomotives, the factor was sometimes as high as 18. In spite of these precautions (sometimes called "factors of ignorance"), ships, bridges, and machinery continued to break.

Toward the end of the nineteenth century the design of warships, in particular, became more competitive and more exacting. The demand for more speed called for longer, lighter hulls. As a result, ships—especially destroyers—were apt to break in two at sea with unacceptable frequency. The British Admiralty finally decided to make actually stress measurements on the hull of a ship at sea in bad weather. In 1903 a real destroyer,

H.M.S. Wolf *tested stresses in bad weather in 1903.*

H.M.S. *Wolf,* was equipped with mechanical strain gauges attached to her sides and keel and decks, and sent to sea to look for bad weather. In a "moderate gale" in the Lizard Race (the sea off Lizard Point, the southernmost point of Great Britain), where the sea was very bad and steep, the highest stress measured was about 12,000 p.s.i. (82 MN/m²). These measured stresses were remarkably close to the theoretical values that had been calculated when the ship was designed. The ship was built from steels having a tested ultimate tensile strength of between 60,000 and 70,000 p.s.i., so the factor of safety appeared to be in the region of 5 or 6, and the Admiralty and naval architects in general expressed a certain amount of self-satisfaction.

This brings us to the question of *stress concentrations*—that is, the weakening effect of local changes in the geometry of a stressed material. Almost any sudden change of section is likely to have some weakening effect, but the introduction of holes, notches, and cracks is particularly bad. This weakening effect must have been known for centuries to craftsmen, especially to those who worked with brittle materials such as stone or glass. However, it seems to have been ignored by nineteenth-century engineers and naval architects, who perhaps assumed that their "advanced" materials, such as wrought iron and steel, were not subject to the same laws.

It is enlightening to plot the *stress trajectories* in a complicated structure. Stress trajectories may be thought of as the paths or lines along which the force or stress may be considered to pass from one molecule to the next through the interior of a solid that is under load. In a uniform, parallel plate or bar subject to an axial tension, the stress trajectories will be a series of parallel lines. If we constrict the plate or the bar in the middle in a smooth, symmetrical way, then the stress trajectories will be crowded closer together in the narrow part. The stress will be higher there, but usually only in proportion to the reduction in cross-sectional area.

However, if we make not a smooth constriction but a sharp notch or crack, then the trajectories will be bunched together at a point of discontinuity in such a way that the local stress may be increased very considerably indeed. One has only to make a scratch or a sharp groove in a piece of long or flat hard candy to see that this is the case: the candy will break along the score. Glaziers use this method to break panes of glass to the desired size and shape.

The effect of a round hole in a uniform elastic material was calculated by G. Kirsch in Germany in 1898. He found that the hole raised the stress in the material in tension by a factor of 3. However, nobody seemed to be much interested. A few years later the problem of the effect of holes and cracks was attacked on a broader front by G. V. Kolosoff on behalf of the Imperial Russian Admiralty. At the time the Russians were much concerned with rebuilding their fleet, which had been largely destroyed by the Japanese in the war of 1905. Kolosoff solved the mathematical problem of the effect of an elliptical hole in a plate and published his results in 1910. Perhaps the tsarist government was somewhat less secretive than their successors. Nevertheless, nobody in the West seems to have paid any attention, possibly because the work was originally published in Russian in St. Petersburg.

The definitive paper on the subject, at least in Western eyes, was published by an Englishman, C. E. Inglis, in 1913. Inglis was at the time an able young academic engineer who later became Professor of Engineering at Cambridge University. Inglis calculated that if an elliptical hole of length $2L$ and tip radius r is subjected to a transverse stress, the stress σ on material at the tip will be raised by a factor of

$$1 + 2\sqrt{\frac{L}{r}}$$

Although this is strictly true only for elliptical holes (which are not particularly common), it is approximately true for holes of other geometries—such as ordinary cracks—and also for things like rectangular hatchways, which are a common cause of spectacular failures in ships.

When Griffith was doing his work on the strength of glass around 1919, Inglis's work on stress concentrations was comparatively new. Griffith proceeded to apply Inglis's ideas to glass. He suggested that solid glass was liable to be full of very fine internal cracks. Although most of these

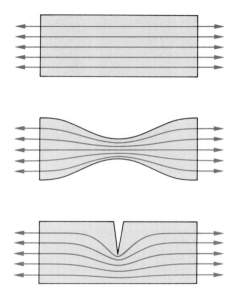

Stress trajectories. Top, in a uniform material subject to axial tension. The stress passes from one molecule to another in a series of parallel lines. Middle, with a smooth constriction. If the cross section of the material is reduced in a smooth or gradual way, the stress will be increased in the narrowed portion in proportion to the reduction in area. Bottom, around a sharp constriction. If the constriction is a sharp one, such as a crack or a notch, the stress trajectories will be crowded together at the crack tip and the local increase in stress at the tip may be very large.

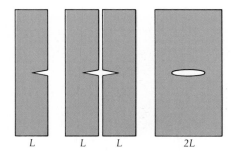

Incidentally, a surface crack may be treated as half an ellipse. In other words, if the crack has a depth L, the stress concentration will again be

$$1 + 2\sqrt{\frac{L}{r}}$$

cracks would have about the same tip radii—that is, approximately molecular dimensions—the cracks in the larger specimens of glass would be longer than those in the very thin specimens. Consequently the stress $\sigma (1 + 2\sqrt{L/r})$ would be greater, and the material would be weaker.

Griffith seems to have supposed that what came to be called *Griffith cracks* were present more or less uniformly throughout the body of the glass, and were due, perhaps, to a failure of the molecules to join up properly when the material cooled. For a number of years after the publication of Griffith's 1920 paper, Griffith cracks were thought of—by the few who thought of them at all—as defects dispersed at random on a molecular scale throughout the body of the material. It was considered likely that they were so small that no one could see or detect them by any of the viewing methods then available.

In 1937, however, E. N. da C. Andrade '"decorated" the surfaces of glass rods with sodium vapor. Using an ordinary optical microscope, the researchers could see that the sodium had condensed in an irregular linear pattern on the surface of the glass, which now looked like a crazy paving. At the time, and for a while afterwards, Andrade's claim that the sodium decoration had revealed the pattern of the Griffith cracks on the surface of the glass was regarded as controversial or at least unproven. There is little doubt that although the hot sodium vapor may have extended the preexisting crack system somewhat, Andrade was right and the rather ill-defined , straggling lines that he saw and photographed were related to real surface cracks on the glass.

Even where Andrade's discovery of cracks on the surface of the glass was admitted, it remained to prove the rest of Griffith's assumption: that there were also many internal cracks throughout the body of the glass. Further evidence came in around 1952, when the British scientist John Morley and others were able to show that if thick rods and fibers of glass were very carefully made and very carefully treated, these, too, would show high strengths. As an important clue to the location of Griffith cracks in ordinary structures, the researchers noted that if the surfaces of the strong glass rods were touched or rubbed, the strength of the rods was much reduced. On the other hand, if the freshly drawn surfaces were protected by a thin layer of metal, such as soft aluminum, the strength was maintained even after moderate abrasion. This was equally true for thin fibers and for thick rods. Thus it became clear that Griffith cracks were a surface phenomenon probably due mostly to accidental damage.

Around 1956, Margaret Parratt, David Marsh, and I, working at Tube Investments' laboratories, near Cambridge, spent much time examining the surfaces of glass and other brittle solids. By refining Andrade's sodium technique, Parratt was able to produce some very beautiful photographs of crack patterns on the surface of glass. Many of these cracks were undoubt-

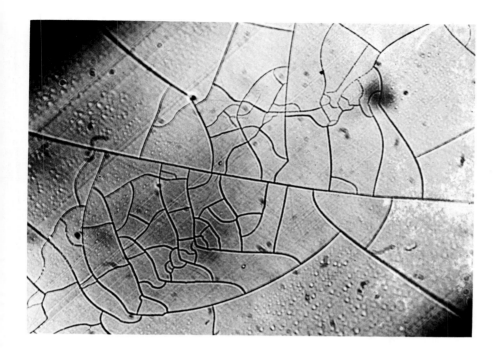

Surface cracks on fresh ordinary Pyrex glass. Such cracks are present on the surface of virtually all glasses and account for their comparative weaknesses.

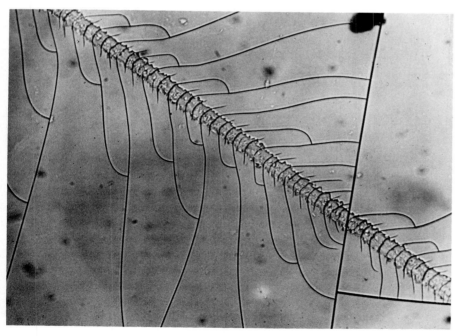

Griffith cracks. Very slight accidental abrasion to glass is capable of forming surface cracks that can have a serious weakening effect.

TENSILE STRENGTH AND TENSILE FAILURE

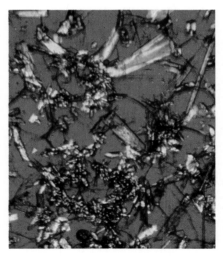

Above, fourth-century Roman vase with decorative doves. Below, cracks are sometimes produced spontaneously on the surface of glass by devitrification—local crystallization of the material. Ancient glass is often extensively devitrified.

edly due to mechanical abrasion. Even the lightest contact with another solid will usually damage the initially smooth surface of a hard material like glass in such a way as to cause very serious loss of strength. In the case of glass, however, there is a subsidiary cause of weakening: fine-scale local crystallization or *devitrification*.

Morphologically, glass is not a crystalline solid but rather a super-cooled liquid that retains the random molecular arrangement characteristic of liquids. Under certain circumstances the molecules in a glass will crystallize because this is a more stable and energetically probable state. When crystallization does occur the consequent molecular rearrangement causes an increase in density and thus a local contraction of the material. This contraction is often sufficient to cause the newly formed crystal to crack. Ancient (e.g., Roman) glass that has survived to the present day is often extensively devitrified, and thus it is very weak mechanically.

By examining modern glass in the electron microscope, David Marsh could sometimes find very small crystals that had cracked as they formed. He was able to show that these cracks sometimes spread from the crystals into the surrounding glass, so they must have had a significant weakening effect. Although devitrification is probably not a very common cause of Griffith cracks, Marsh did show that weakening defects can sometimes arise spontaneously and are not necessarily due to external causes such as abrasion.

So far we have been discussing glassy materials; what about crystalline ones? Most metals are crystalline and many of them are *ductile*—that is, they flow fairly readily when subjected to shear. Such metals have special failure mechanisms that we shall discuss in Chapter 5. Very hard metals and nearly all nonmetallic crystals do not flow in this way but break under tension in a manner that is not very different from glass.

This brings us to the subject of *whiskers*. When most substances crystallize they form thick, bulky crystals, often cubes or similar shapes. However, for reasons that we shall examine in the next chapter, it is possible for many crystalline materials to grow from solution or from vapor in the form of long, thin, hairlike crystals known as whiskers. Metal whiskers have had an unfavorable reputation within the electronics fraternity for many years. These pesky crystals will sometimes grow, rather like fungus, on metal surfaces inside electrical devices, causing short circuits, bad language, and sometimes considerable expense.

In 1952, two American researchers, C. Herring and J. K. Galt, found that whisker crystals of metallic tin, which is normally a soft, weak material, exhibited elastic strains of about 2.0 percent, which corresponds to a stress of around 150,000 p.s.i. or 1,000 MN/m^2. Not long afterwards, I grew whisker crystals of hydroquinone and other water-soluble nonmetallic sub-

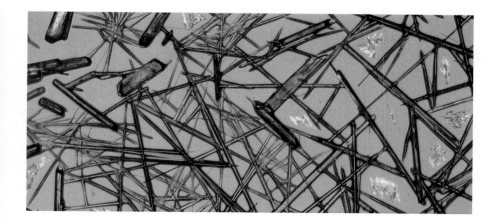

Whisker crystals. Although the surfaces of the original thin whisker crystals are smooth, whiskers thicken by the spreading of "growth layers" that have a sharp or stepped front.

stances and found that they, too, showed very high strengths. However, since the only testing method available at that time was bending the crystals on the stage of an optical microscope, these results were very rough.

A bit later, about 1956, David Marsh came to work with me. Marsh was able to construct a microtesting machine that could subject very small fibers and whisker crystals to controlled tension. Marsh was able to grow and test whisker crystals of a large number of substances. Most of these filaments showed a strength-to-diameter relationship that was remarkably similar to the one Griffith had found for glass. In fact, graphs of the effect of size upon elastic breaking strain revealed that most of the materials plotted on the same curve.

Because of the manner in which these crystals were known to grow, it was unlikely that the weakness of the thicker ones was due to Griffith cracks. However, other geometrical irregularities besides cracks are capable of producing severe concentrations of stress. When a whisker crystal first forms, it generally grows by means of a *screw dislocation*—a mechanism we shall examine in Chapter 5. As a result, the original thin fiber is very smooth indeed. Subsequent thickening of the fiber occurs by the spreading of "growth layers" along its surface, which results in the formation of sharp steps. The thicker the crystal becomes, the higher these steps are apt to be: if a thick needle crystal is examined under the microscope, it is usually seen to be covered with prominent steps rather like miniature cliffs.

Working with photoelastic models, Marsh was able to show that the stress concentration at the point where a step joins the substrate (that is, at the bottom of the "cliff") is nearly equal to that at the tip of the equivalent

Photoelastic pattern of sheer stresses around the root of a step.

crack. (This experimental work was confirmed not long afterwards by a mathematical analysis made by the British physicist H. L. Cox.) Marsh also observed that the average radius of the notch at the bottom of the step remained about constant at around 40 angstroms. Because the thicker crystals had higher steps, the stress concentration at the notch was worse in the thicker crystals. The measured strengths of a considerable range of crystals correlated pretty well with step heights. In other words, these steps have much the same weakening effect as do surface cracks in materials like glass.

In an attempt to explain why brittle solids fail to exhibit their theoretical tensile strengths, the force argument has undergone some changes since Griffith first proposed that tiny internal cracks were responsible. Subsequent scientists (Morley et al.) refined Griffith's concept from widely dispersed internal cracks to surface cracks resulting from very slight abrasions—although there is some evidence that, for glass, at least, internal cracks can arise spontaneously. A further elaboration of the causes of surface weakness came with Marsh and Gordon's discovery that for crystalline substances, uneven steplike surfaces result from the normal growth of crystals. Surface flaws are only half of Griffith's story, though: he had other weapons in his arsenal to deal with the phenomenon of lower-than-theoretical strengths under tension.

Photoelasticity. Polarized light has its plane of polarization rotated when it is passed through a transparent solid which is under stress. The amount of rotation is proportional to the stress. This method is commonly used for stress analysis using transparent models (e.g., of Lucite). Robert Mark's analysis of stress in medieval cathedrals (page 14) and in Wren's Sheldonian Theatre (page 17) uses this method.

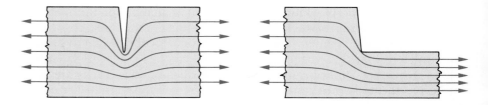

THE ENERGY ARGUMENT: MODERN FRACTURE MECHANICS

Inglis's stress concentrations were accepted for many years as an adequate and reasonable explanation for the general failure of solids to reach their theoretical strengths. However, as Griffith pointed out, the force or stress concentration argument was, in a way, too good: according to this explanation, many solids ought to be much weaker than they actually are.

If we suppose, as Griffith did, that the effective diameter of the tip of a Griffith crack approximates the intermolecular spacing, which is, say, about 2 angstroms (2×10^{-10} m) then a surface crack with a depth of 1 micrometer (10^{-6} m) would produce a stress-concentration factor of around 200. This value is on the high side, but it is roughly in line with the observed strength of bulk glass. However, a crack 1 centimeter (10^{-2} m) deep would reduce the theoretical or intermolecular strength by a factor of 20,000—in which case ordinary windows, which often have cracks this long, could scarcely survive. Large structures such as ships often contain cracks at least 1 meter long, which would produce a stress-concentration factor of 200,000, indicating that the structure would not be able to support its own weight.

The assumption behind the force or stress-concentration argument is that if at *any point* within a solid the local stress is sufficient to separate neighboring molecules, then that separation or fracture will spread right across the material and cause it to break into two or more pieces. But it is an imperfect world, and in real materials—both crystalline and amorphous—all the molecules do not necessarily join up perfectly when the material solidifies. There are likely to be local defects where a few molecules are too far apart to attract each other properly. Even though this is probably the case, to some extent, in most solids, such materials seldom fall to pieces spontaneously, and many of them are very strong.

Furthermore, if we consider a large, elderly structure—say, a ship, a bridge, or an automobile—it is likely to be full of holes, notches, cracks, scratches, corrosion pits, and so on. If we were to calculate the Inglis stress concentration at the worst of these defects we would probably get a fright.

Growth of a crack on a carbide specimen. The general direction of growth is from left to right. The tip of the precrack is the dark area at the far left. Just to the right of this (above the label A) are microvoids formed in the immediate vicinity of the tip. Farther right (above the label B) is a grain of carbide with a microcrack running across it.

An engine is a mechanism for propelling an automobile, but it will only do so if it is kept supplied with fuel. A saw is a mechanism for cutting wood but it needs to be supplied with muscular or mechanical energy from some external source. If we get tired or switch the power off, the cutting operation will cease.

But with a little good fortune, such structures go on carrying their loads with comparative safety for years, sometimes for hundreds of years.

Griffith's most important statement was that the force or stress-concentration argument is not adequate *by itself* to account for the fracture of a material. It is a necessary, but not a sufficient condition. In effect, a stress concentration is only a *mechanism* for separating atoms or molecules and enabling fracture to take place. Like any other mechanism—e.g., a saw or an automobile engine—if it is to work it needs to be fed with the right kind of energy.

A material or a structure that is carrying a load is, in effect, a spring: a store or reservoir of strain energy. This strain energy is diffused throughout a considerable volume of material. If fracture is to take place, a sufficient quantity of this diffused energy must be transmitted to the fracture zone to operate the fracture mechanism.

Consider a piece of material such as that shown in the figure below. In the original unstressed state (left), it will, of course, be without any strain energy. If we subject it to a tensile stress σ, the material will now contain $\sigma^2/2\,E$ units of energy per unit volume. If we clamp the material rigidly at both ends (center), no mechanical energy can get into or out of the system. In the undamaged material this energy may be considered to be distributed more or less uniformly. Now suppose that a crack of length L is introduced into the system (right). The material on either side of the crack—that is, the shaded area in the diagram—can now relax, because the crack will "gape" to some extent. In relaxing it will release most of its strain energy, and this released strain energy becomes available to operate the fracture mechanism and so cause the crack to spread.

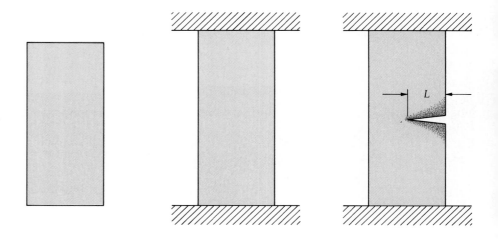

Strain and cracks. At left, unstrained material. Center, material strained and rigidly clamped; no energy can get in or out of the system. At right, clamped material is cracked; the colored areas around the crack relax and give up strain energy, which is now available to propagate the crack still further.

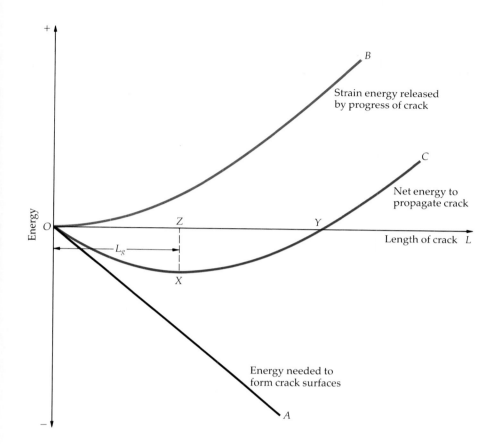

Strain energy released
by progress of crack

B

C

Net energy to
propagate crack

Length of crack L

Energy needed to
form crack surfaces

A

Release of strain energy (see discussion below and on pages 27–29).

As we said at the beginning of the chapter, the creation of new surfaces during fracture requires energy—*at least* as much per surface as the free surface energy γ. In practice, the making of a new fracture surface may need considerably more work than is accounted for by the free surface energy. Let us use W to designate the *work of fracture,* the total amount of energy required to expose the two new surfaces resulting from a fracture. Then the total amount of energy needed to create a new crack of length L will be $2WL$. In other words, the work consumed in causing a fracture bears a *linear* correspondence to the length of the new crack.

On the other hand, the strain energy *released* by the system during the fracture will usually be proportional to L^2, represented by the area of the shaded triangles at the right in the figure on page 88. Thus the curve of strain energy released is usually a parabola. Adding the curves for the consumption and release of strain energy together (see the figure on page 89), we find that up to some point X, the whole system as a whole is *consuming* energy, so that the crack has no energy incentive to propagate. Beyond point X the system begins to *release* energy, so that the crack will propagate of its own accord. It follows that cracks of less than the *critical Griffith length L_g*, which is equal to OZ, are energetically stable and thus usually safe; cracks of greater than this critical length are energetically unstable and will usually propagate faster and faster.

Although Griffith's mathematics are somewhat daunting, the finished result is really very simple:

$$L_g = \frac{1}{\pi}\left(\frac{\text{work of fracture per unit area of crack surface}}{\text{strain energy stored per unit volume of material}}\right)$$

Putting this algebraically, we have

$$L_g = \frac{2WE}{\pi\sigma^2}$$

Note that Young's modulus is given in newtons, *not* meganewtons.

where W is the work of fracture (in joules per square meter), E is Young's modulus (in newtons per square meter), σ is the average tensile strength (taking no account of stress concentrations) in the material near the crack (in newtons per square meter), and L_g is the critical Griffith crack length in meters.

If we equate the work of fracture W to the surface energy γ of a solid, simple algebra shows that there is little difference between the force and the energy approach to the problem: a defect of length L will have much the same effect in either case. However, Griffith pointed out that even in the case of very brittle materials such as glass, the actual amount of work needed to cause fracture is higher than the resulting surface energies. This is because not only do the atoms and molecules have to be separated in the plane in which complete fracture occurs, but the whole of the fine structure of the material is sometimes disturbed to a depth of a centimeter or more below the actual fracture plane; in other words, interatomic bonds may be broken in regions that are half a million atoms away from the fracture plane. The total energy needed to break all these more distant bonds can be very large indeed.

Even in very brittle materials like glass and pottery the practical work of fracture may be around 10 times the actual free surface energy. In fact, the ease with which glass can be indented at room temperature in the laboratory indicates that there may always be a certain amount of distortion of the molecular structure in regions near the surface. Even such minor localized distortions absorb a measurable amount of energy and make the glass slightly tougher than it would be otherwise.

If a material has a really large work of fracture, as does steel or wood, a specific work of fracture mechanism must be operating. In other words, there must be some mechanism by which energy can be absorbed in large quantities in regions of the material away from the fracture surfaces. In tough metals, the mechanisms seem to be, in a sense, accidental (rather than purposive, as in biological and synthetic materials). In tough biological materials such as wood, the work-of-fracture mechanisms are very sophisticated. In any case, whether a good mechanism just happens as a consequence of the nature of the metallic bond or whether it has evolved by some Darwinian process, it may be very effective indeed: some mechanisms are able to increase the amount of energy needed to form a given length of crack by factors of well over 100,000 as compared with the simple free surface energy. Because the critical crack length L_g that will cause fracture at a given stress in tension is directly proportional to this work of fracture W, the value of W and the nature of the mechanisms on which it depends are of the greatest importance in both engineering and biology. Some values for a few common solids are given in the following table.

Work of fracture of some common materials

Material	Approx. free surface energy (J/m^2)	Approx. work of fracture (J/m^2)	Approx. bulk tensile strength (MN/m^2)
Mild steel	2.0	500,000	400
High tensile steel	2.0	10,000	1,000
Aluminum alloys	0.9	100,000	400
Wood	0.1	15,000	100
Bone	0.2	1,700	120
Glass and pottery	0.5	1–10	170

Note that the normally accepted "tensile strength"—determined by breaking an ordinary smooth engineering test piece, and such as is listed in the handbooks—has a very irregular connection with the work of fracture. In metals the relationship is a reverse one: the "weak" low-tensile steels are tough, whereas the "strong" high-tensile steels are brittle.

Here the scale of a component or a structure becomes important. The critical Griffith crack length L_g is an *absolute*, not a relative, distance. For the same material at the same stress, the values will be identical in a wristwatch and a battleship. This is where the Griffith doctrine begins to bite.

With fairly small objects, such as swords or crankshafts, the longest dangerous crack is unlikely to be more than a few millimeters long. It is therefore often safe and practicable to use a strong, hard, high-tensile steel with a moderately low work of fracture for small components.

When we come to larger structures, the dimensions both of deliberate openings, such as doors and hatchways, and of accidental cracks must be proportioned to the calculated critical crack length. With metals, this will mean that an alloy of high ductility—that is, a high work of fracture—will be needed. These tough, low-tensile alloys need to be worked at a low stress to ensure an acceptable critical crack length.

In ordinary mild steel, such as that used in the construction of bridges and ships, the critical crack length at a working stress of 11,000 p.s.i. (80 MN/m^2) is about 2 meters. If the working stress is raised to, say, 15,000 p.s.i. (110 MN/m^2), the critical crack length will be reduced to 1 meter. It follows that mild steel is like to be a safe material for structures of moderate size, but in really large structures there are nearly always panels and other parts that are much more than 1 or 2 meters across. In fact, a considerable number of the accidents in recent years to bridges, ships, and aircraft can be at least partly attributed to the failure of the designer to utilize modern ideas about fracture mechanics—that is, the energy approach to fracture.

For many years engineers—safety authorities as well as designers—have been conditioned to think in *force* terms. Deeply ingrained in the traditional engineering mind is the idea that every material has a "safe working stress" that is documented in the literature: if this stress is not exceeded, then the structure will supposedly be safe. The Griffith concept of a structure as an energy system that is potentially unstable—as a store of energy that is trying to escape—has been resisted with great intensity.

A loaded structure is actually closely analogous to a chemical explosive. A "safe" explosive is not detonated by a very local increase in temperature, such as might occur from a slight mechanical impact. If only a small volume of the explosive material is heated, the rate of loss of heat to the surroundings will be greater than the supply of heat from the chemical decomposition of the explosive, so the explosive reaction will not spread. Thus a *critical volume* of an explosive material needs to be activated before the energy of the explosive charge, as a whole, is released. The business of the explosives experts is to ensure not only that the chemical energy stored in their materials is high, but also that it can be released only under predictable conditions.

Protection against mechanical failure is comparable in many ways to protection against a chemical reaction such as combustion or explosion. In both cases the release of a large quantity of stored energy is prevented by an energy barrier that is linked with a scale or dimensional effect. In the

THE DANGER OF SCALING UP

Engineering design practice is based to a large extent upon experience and precedent. One successful project leads to another, and things generally get bigger and bigger. The tendency is to scale up a successful design, keeping the working stresses the same. Not very long ago the author had a head-on collision over the conference table with NASA about its wind-energy program. A number of safe and successful windmills are 175 feet in diameter. NASA proposed, in effect, to scale up this design to a diameter of 400 feet. I had to point out that, for reasons of modern fracture mechanics, extensive redesign would be needed if the bigger windmills were to operate safely. The atmosphere became distinctly tense and it became clear that the doctrines of Griffith were not fully understood or accepted in certain quarters. (Fortunately, I have reason to believe that the design of the big windmills has since been modified.)

A similar but much more serious failure to come to terms with modern fracture mechanics occurred in England around 1970. The Rolls Royce Company invested a great deal of time, money, and effort in the development of a new jet engine, the RB211. The design and performance of this engine depended to a large extent upon the use of carbon-fiber fan blades. Some of us who had knowledge of the project wrote a minority report pointing out that the work of fracture of the carbon-fiber blade material was inadequate to cope with the impact of, for instance, small birds. This report was ignored by the powers that be. The carbon-fiber blades did prove to be too brittle, and as a direct consequence the RB211 engine had to be redesigned with metal blades and Rolls Royce went bankrupt in 1971. Fortunately, nobody was killed as a result of the failure of these blades, but the financial loss was in the region of 2 billion pounds sterling (about 4 billion dollars at the time).

The NASA MOD-OA wind turbine generator at Kahuku Point, Oahu, Hawaii, employed wood epoxy blades. Erected in the late 1970s the machine's blades measured 125 feet in diameter and weighed 2500 lb. The 400-ft-diameter wind turbine generator the author advised NASA on was never built.

case of a stressed material, the important dimension is the length of the critical Griffith crack.

At the stresses at which most practical structures are operated, the amount of strain energy stored per unit mass is far less than the chemical energy stored in explosives. Consequently an engineering test piece usually breaks into two halves in the testing machine with only a moderate bang. Even so, in highly stressed structures such as aircraft, the quantity of energy stored is far from negligible. At Farnborough it is, or used to be, the practice to test complete large aircraft to destruction in the test frame. Such tests required a considerable number of assistants to read the various dials and gauges. When the structure was obviously nearing failure, the officer in charge of the test would shout a warning and we would all run for cover. The final failure of a large airframe can be a dramatic business once the critical stage is passed.

Now that it is easier to achieve in the laboratory close approximations to the theoretical strengths of many solids, we have to face the fact that the strain energy stored in a material under a load approaching its theoretical capacity approximates to the total chemical bond energy. In fact, the strain energy released at fracture by these very strong specimens is roughly the same per unit weight as that released by a conventional chemical explosive such as dynamite.

Thus when very strong fibers and crystals are broken in devices such as Marsh's microtesting machine, the result is quite different from that observed when ordinary engineering test pieces are broken. Instead of a mild bang and two broken pieces, there is an explosion followed by a little cloud of dust or smoke. Otherwise there are no visible remains of the test specimen, which has simply vanished. Only the very small size of such test specimens prevents them from doing serious damage to both the equipment and the experimenter.

Clearly, attempts to subject materials to stresses near the theoretical maximum in a real structure would be far too dangerous: one might as well try to build an aircraft out of dynamite. This does not mean that the stresses in engineering structures cannot be raised above their present levels, but to raise them even moderately with safety requires considerable skill and knowledge on the part of the designer.

CRACK STOPPING IN TENSION

The absolute nature of the critical Griffith crack length favors small structures, which consequently tend to be stronger than larger ones. There is no need to plan for defects a meter long in a structure that is less than a centimeter across. Small plants and animals are much more successful—and hence more numerous—than large ones. Insects infest the whole earth, but elephants are rare outside zoos. Grasses are more common are more persistent than large plants. However, large plants such as trees are much more numerous, generally bigger, and longer lived than large animals. We shall discuss the structural reasons for the enormous success of timber as a living material in Chapter 6.

If we have to plan only for cracks that might be a millimeter or, at most, a centimeter long, then we can generally work with safety at higher stresses and make use of different and sometimes more effective materials. In engineering these conditions are applicable to things like the small parts of machinery and small springs. The greater safety of small structures is one reason why it is beneficial to subdivide materials. Wire rope is a case in point. It can be safely made from a brittle, high-tensile steel because the steel is subdivided into many strands.

In a rope or a cloth, if one strand breaks in tension the fracture will not, as a rule, spread to the neighboring strands. This is because there is no means by which the released strain energy can be transmitted across the gap between the fibers. When a material is breaking, the released strain energy is transmitted to the fracture zone by an elastic mechanism involving shear. If there is no effective adhesion between the fibers or other component parts of the material, the fibers will simply slide over each other and the fracture process will stop. If we saturate ordinary organic rope or cloth with an adhesive that hardens and becomes rigid, the rope or cloth will become brittle and much weaker because the rigid glue is able to transmit considerable shearing forces and, with them, enough strain energy to keep the fracture process going.

If we use a flexible adhesive, such as rubber, the rubber will be able to transmit very little shear and the strength of the rope or cloth will be largely preserved. This is why various rubberized fabrics are used for waterproof garments, pneumatic boats, and so on. Similar materials include tendon, in which thin, parallel, stiff filaments that transmit the forces from our muscles to our bones are separated from one another in this kind of way. Tendon is an exceptionally strong and tough material. It is capable of storing safely about 20 times as much strain energy, for its weight, as high-tensile steel. For this reason animal tendon was widely used in the ancient world for storing the necessary strain energy in weapons like bows and catapults. This biological material may have some important lessons to offer modern engineers.

Most present-day engineering structures are made from conventional metal alloys. Safety is achieved by subdivision in one way or another. Lattice structures, for instance, are much safer than shells or monocoques. For many years American battleships had lattice masts. During World War II, the "geodetic" bomber aircraft, such as the Wellington, had a fabric-covered metal lattice airframe and an enviable safety record. Wellington

Subdivision into many thin strands accounts for the strength of wire rope.

Wellington bombers—the Wimpey of World War II—under construction.

bombers frequently made it home from raids on Germany after suffering damage from antiaircraft fire that would undoubtedly have destroyed an aircraft with a continuous metal skin. Because of its structural safety the Wellington was much loved by the Royal Air Force and generally known as the Wimpey.

The fabric covering of the geodetic aircraft was not smooth and the metal lattice was heavy, so its design was unacceptable for modern pressurized airplanes. Ships and nearly all modern aircraft have a continuous metal skin. Where the joints in the skin plating are riveted, the discontinuity provides some degree of subdivision and has at least a partial crack-stopping effect. However, when the joints in the plating are welded, as they nearly always are in modern ships, a crack is free to run as far as it chooses—quite often right across the ship. Local thickenings such as ribs and stringers do have a certain crack-stopping effect, but they are not entirely reliable.

Even so small a monocoque as the human skull is provided with *sutures*, anti-Griffith devices resembling dovetailed joints. Ordinary eggshells are designed as monocoques to make it easy for the young birds to peck their way out. Lobsters and similar animals do have tough and durable shell structures, but Nature has been to a lot of trouble to raise the work of

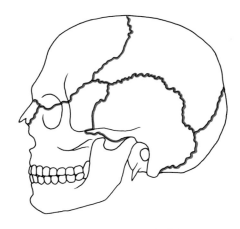

Sutures in the skull are anti-Griffith devices.

fracture of lobster shells. Even so, Griffith considerations set a strict limit on the size of such animals, and their dimensions are below those of ships and aircraft.

An idea current among laymen—encouraged perhaps by pictures in the science fiction literature—is that the ideal structure of the future will be an enormous continuous smooth shell or monocoque, like some vast streamlined egg. Such a concept may be a science fiction writer's dream, but it would be an engineer's nightmare.

One of the main arguments for monocoque construction in aircraft wings and fuselages is the need to provide very high torsional stiffness to prevent such lethal conditions as wing flutter and aileron reversal. (An aileron is a movable portion of the wing.) In flying animals, however, Nature manages differently. Birds are quite flexible in torsion, but they somehow maintain stability.

HAVING IT BOTH WAYS—COOK AND GORDON

Thus far we have been considering materials and structures designed primarily to resist tensile loads. But many materials and structures are called upon to resist both tension and compression. This is especially true of trees and bones, which may be bent in any direction. It is also true of a large number of technological materials.

If a structure is designed only or mainly for tension, there is no weight penalty for subdividing it and there are considerable safety benefits in doing so. This strategy is appropriate for, say, a rope, in which many strands are loosely attached to one another, but such structures are ineffective in compression because the various parts will buckle separately. An efficient strut must be elastically continuous right across its thickness. Bricks and cement, for instance, are fairly efficient in compression because they are homogeneous; but just for this reason they are weak and brittle in tension since cracks can spread across them easily. What makes materials like wood and bone able to "have it both ways"?

Around 1962 John Cook and I decided to examine the conditions at the tip of a crack in more detail than had been done before. The basic mathematics of the stress around an elliptical crack had, of course, been pioneered by Inglis in 1913. Inglis had calculated the rise in stress at the crack tip. This was long before the days of computers, and the enormous amount of human labor required for the arithmetical calculations precluded for many years a more general exploration of the stress field surrounding the crack tip. By the 1960s, though, computers were becoming available, and

John Cook was able to evaluate Inglis's equations with the help of what was then a "state-of-the-art" British machine, the Mercury computer at Farnborough.

The left-hand figure below shows the distribution of the stresses *parallel* to the remotely applied stress—that is, the stresses perpendicular to the plane of the crack. The right-hand figure shows the stresses *in the plane of the crack*—that is, the stresses at right angles to the remotely applied stress. Not entirely to our surprise (since we were looking for it), there is another stress concentration, not at the actual crack tip, but a little way ahead of it. Only this time the concentrated stresses are parallel to the plane of the crack. In a homogeneous material this stress concentration always has a value equal to one-fifth of the Inglis stress concentration factor

$$\frac{1}{5}\left(1 + 2\sqrt{\frac{L}{r}}\right)$$

The Cook-Gordon mechanism for stopping cracks. If the material contains planes of weakness in a direction that is perpendicular or oblique to an advancing crack, the production of a secondary crack may act as a crack-stopper. This is one reason why materials like wood are so tough.

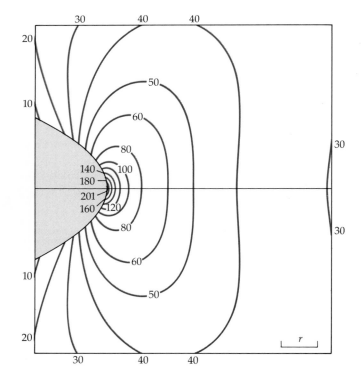

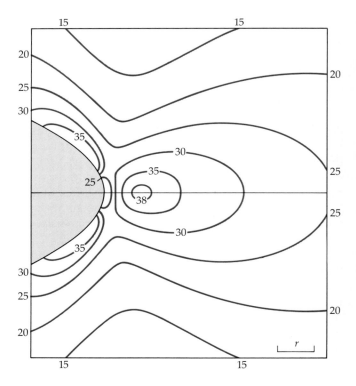

If the material is homogeneous and continuous, this secondary stress concentration has little or no effect upon the mechanism of failure in tension. However, should the material contain plans of weakness—in other words, if it is able to split—a secondary crack may open up ahead of the main crack and act as a crack stopper. Thus we have the apparent paradox that, by putting planes of weakness into an otherwise homogenous material, we can increase its tensile strength considerably.

A striking example of this effect occurs, for instance, with mica. The type of mica known as muscovite cleaves readily. It has a tensile strength in the plane of the laminae of about 460,000 p.s.i. (3,100 MN/m^2), a very high figure that is close to its theoretical strength. Another variety of mica, margarite, is very similar to muscovite except that it has twice the electrical charge across its planes of cleavage: margarite, however, has a low tensile strength. Similar effects are observed with asbestos, another crystalline substance that contains planes of cleavage.

In the case of minerals such as mica and asbestos, the strengthening effect of the planes of cleavage is fortuitous. In biological materials, the introduction of planes of weakness is clearly purposive, widespread, and very effective. It is well known, for instance, that wood and other plant tissues split very easily. This apparent weakness has an important protec-

Photograph of mica.

tive effect: a crack is almost never able to proceed directly across the trunk of a tree. Wood seldom breaks in direct tension; when it does, the fracture is a jagged zigzag, showing that the crack has been repeatedly stopped and deflected.

The Cook-Gordon effect—that is, the strengthening effect of planes of weakness—is also present in bone, where it is less conspicuous. But in fact the bones of humans and other animals are fairly full of incipient cracks that have been halted by secondary cracks. These "stopped" cracks generally heal naturally without our being aware of their existence. Usually the main or primary crack heals first, the secondary crack a little later. In older people, however, the primary crack may heal reasonably quickly but the secondary cracks may not heal at all. Unhealed secondary cracks are apparently quite common in the bones of old people, to whom they are generally harmless.

Secondary cracks are involved in a curious condition that affects deep sea divers. To avoid the "bends"—the absorption of nitrogen by the body tissues during a too-rapid ascent to the surface—divers sometimes breathe a mixture of oxygen with helium, rather than nitrogen. The helium is not absorbed to any extent by the tissues. However, if the diver's bones contain secondary cracks of the sort we have been discussing, helium will be lodged, and perhaps sealed off, within these secondary cracks. The continued presence of the helium not only inhibits the healing of the secondary cracks but is liable to lead to local bone infections that can be troublesome and sometimes dangerous. However, the habit of breathing helium might be described as an unnatural hazard.

The provision of planes of weakness is universal in wood and bone and very common in soft tissues, as anybody who has ever eaten a steak must have noticed. Planes of weakness are common in the shells of oysters and snails. If one watches a bird trying to break the shell of a snail by hammering it upon a stone, one can see how effective the idea is.

In spite of these conspicuous real-life examples of effective crack stopping, it took us a long time to realize that the way to make an acceptably tough artificial composite material is *not* to increase the adhesion between the fibers and the resin matrix beyond a certain point. The earlier composite materials got a reputation for unreliability and brittleness because the adhesion of the resin was too good.

Although Griffith's work at Farnborough on the strength of glass was begun with proper official consent, few of his colleagues and superiors understood or appreciated his results—or respected his work. However, Griffith and Lockspeiser continued to devote a good deal of time to it, more than the powers that be realized. Lockspeiser told me that the two of them kept fairly quiet about it. Unfortunately, Lockspeiser forgot one evening to turn off the gas torch he used for melting the glass. After the enquiry into the resulting fire, he and Griffith were told to stop wasting their time and the work on glass was ended.

After a transfer to the Engines Department at Farnborough in 1926, Griffith began work on development of jet engines. Despite his progress, his work was again met with skepticism and was stopped by officialdom in 1930. It was not until 1937 that the jet engine project was revived. Unfortunately, civil service politics entangled Griffith in bitter and involved controversies and he was never able to play the part he might have in a very successful invention.

5

METALS AND DISLOCATIONS

Swords, Plowshares, and Satanic Mills

Ay me! What perils do environ
The man that meddles with cold iron;
What plaguy mischiefs and mishaps
Do dog him still with after-claps!
For though Dame Fortune seem to smile
And leer upon him for a while,
She'll after shew him in the nick
For though Dame Fortune seem to smile
Of all his glories, a dog trick.

SAMUEL BUTLER (1612–1680)
"Hudibras"

As Gilbert Murray pointed out in his fascinating *Five Stages of Greek Religion*, although the gods of most nations claim to have created the world, the Olympian gods merely claimed to have conquered it and to have imposed themselves, as deities, upon it. Their humanlike attributes may reflect a historical parallel: it is likely that the Dorian invaders from the north conquered the Greek world with iron weapons, around 1400 B.C., when the Minoan kingdoms—which had depended on bronze tools and weapons—collapsed. Not surprisingly, the Olympians found room for a smith-god, Hephaistos, who later became the Romans' Vulcan, after whom volcanoes are named. On Mount Olympus, the boardroom of the gods, Hephaistos was, however, kept in his place as a technologist; his business was to supply weapons, not to decide policy. Moreover, he was lame, so that he could not run away and defect to some rival cult or ideology.

Like the Olympian gods, civilized mortals have long regarded metals in general, and iron and steel in particular, with mixed emotions. For one thing, these materials are not "natural": they have nothing to do with plants or animals. They are born of fire—the smith's forge or the great furnaces of modern industry. They are associated with strength and power, with weapons and politics, and with the satanic mills of Detroit and Manchester. In Germany in the nineteenth century, Count Bismarck became the "Iron Chancellor." Later, in Russia, one Joseph Djugashvili changed his name to Stalin: the word for steel in many modern European languages derives from the Teutonic verb *stalen*, which meant *to harden* or *to steel*. Former United States First Lady Rosalyn Carter was dubbed the "Steel Magnolia" in tribute to her soft-spoken determination; in modern England, journalists refer to Prime Minister Margaret Thatcher as the "Iron Lady."

For 4,000 years metal tools and weapons—axes and horseshoes, swords and guns—have played an enormously important part in history. However, the quantity of metal used for these purposes was comparatively small. The really massive use of metals, especially steel, has been for large structures—ships and bridges, vehicles and heavy machinery—built in the last 150 years. Although it seems likely that metals will go on being used in most of their traditional roles—e.g., for weapons and cutting tools—for a long time to come, steel is not a particularly efficient material for large structures, and the long-term future of the heavy steel industry seems open to doubt.

Metals are common in the earth's crust, where they nearly always occur in the form of chemical compounds such as oxides and carbonates. In the uncombined elemental state they are rare, although, judging from the composition of many meteorites, uncombined metals such as iron must be fairly common in outer space. In biology, although compounds of iron and of calcium, for instance, play important roles in plants and animals, I do

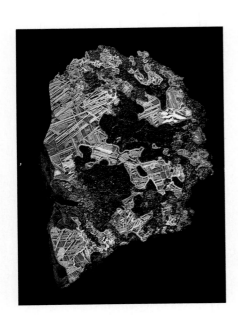

Uncombined metals are more common in space than on the surface of the Earth—at least from the evidence of meteorites.

not think that there is any instance of the use of metals, as such, for anything that can be described as a structure. The exploitation of metals to carry stresses seems to have been entirely a human achievement.

Unlike flint or bone—the materials used by the earliest humans—metals could be forged, cast, or cold-worked into complicated shapes. Not only were they comparatively strong and rigid but they could be made tough, or else they could be hardened and ground to a sharp cutting edge. Moreover, unlike most biological materials, metals are *isotropic*—that is, they have almost the same mechanical properties in every direction. Although copper and tin have always been fairly scarce, iron ore is plentiful. Once means of reducing the ore to usable metallic forms were developed, iron and then steel became immensely important.

Yet science came very late to the world of metals. Even the simple chemistry governing the extraction of metals from their ores was not well understood until the middle of the nineteenth century. Once the metal was extracted, metallurgists and engineers knew that they could add this ingredient or that, or that they could heat, cool, or hammer the metal according to traditional procedures, to change its mechanical properties, sometimes very dramatically. Although this knowledge was quantitative—so many percent of carbon would change the tensile strength by so much—the whole subject remained, like cookery, largely a pragmatic one. By the 1930s a great deal was known about relativity and astronomy, and I remember hearing Lord Rutherford lecturing on the structure of the atom, but almost nothing was known in any fundamental molecular sense about why the various metals used in making different parts of an automobile behaved in their various different ways. Lecturers in metallurgy tended to avoid the subject. In those days, questions of this kind were regarded not merely as tactless, but as interdisciplinary and therefore in rather bad taste.

DISLOCATIONS AND G. I. TAYLOR

Although very hard, brittle metals often break by stress-concentration mechanisms that are substantially the same as those Griffith identified in glass, the important mechanical characteristic is their *ductility:* their ability to yield without breaking when subjected to shear (lateral stress causing parallel planes of atoms to shift in relation to one another). This shearing usually occurs on many planes that are generally about 45° to the direction of the applied stress. The result is that the metal behaves in a plastic manner a little like a very strong, tough clay.

How do we account for the ductility of most fairly pure metals? Until 1934 no convincing explanation existed. In that year, the Cambridge engineer Sir Geoffrey Taylor put forth the hypothesis of *dislocations* in crystals.

Ductile materials such as metals fail in a plastic manner by shearing, usually on many planes that are roughly at 45° to the applied tensile force.

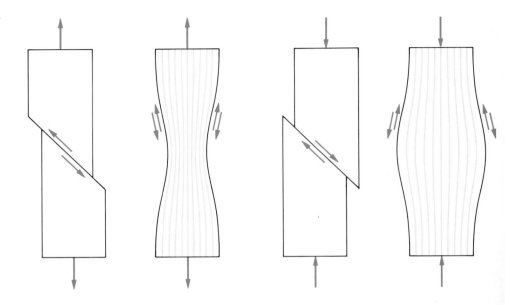

Taylor argued that the ductility of metals could not arise from point defects in the crystal structure: for example, holes or extra atoms would be more likely to produce Inglis-type stress concentrations and cracking. The cause of plasticity would have to be a *line defect* running across the three-dimensional structure of the crystal. This would enable whole planes of atoms to slide over each other and advance, like an army, on a broad front.

Usually the metal is not weakened by this plastic distortion; very often it becomes stronger; this is known as work hardening. Plastic behavior of this sort generally relieves the concentrations of stress which are only too apt to occur at holes and cracks in the material. It also enables the metal to be shaped to a certain extent at room temperature by bending, stretching, or hammering.

However, if we distort a metal too much by cold-working, it will become brittle and eventually break. But a cold-worked metal can be restored to its original ductile state by annealing it; that is, by raising the metal to a moderately high temperature and then allowing it to cool slowly. The process of cold-working and annealing can be repeated almost indefinitely, as the ancient armorers discovered.

The most important technological characteristic of metals is their ductility, their ability to yield without breaking under shear forces: on the one hand, this ductility enables them to be shaped; on the other, it makes them tough. The stress at which metals will yield under shear can vary enormously. Pure metals such as gold, silver, and lead are very soft indeed. So is the aluminum that is made into foil used for packaging. Even pure iron is fairly soft. However, the addition of small amounts of other elements—of tin to copper to make bronze, or of carbon to iron to make steel—can considerably increase the hardness and strength over that of the pure metal.

Some approximations to the theoretical shearing strength of various metals at room temperature calculated by Anthony Kelly are shown in the following table.

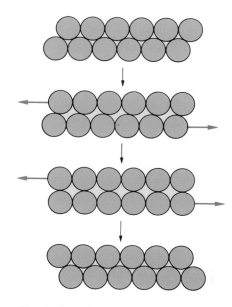

Shearing by sliding of whole planes of atoms without benefit of dislocation mechanisms. At top, the initial position at rest. Second from top maximum resistance to slip, with about 15° angular shearing displacement. Third from top, all resistance to shearing is gone at 30°. Bottom, final position of rest at 60° shear.

Theoretical Shearing Strengths of Various Metals

Metal	Shearing Strengths	
	$MN/m^2 \times 10^4$	$p.s.i. \times 10^6$
Iron	0.68	1.0
Copper	0.12	0.175
Aluminum	0.09	0.13
Silver	0.08	0.11
Gold	0.08	0.11

Although these values for theoretical shearing strengths are appreciably lower than those for theoretical tensile strengths, they are still very much higher than the shearing strengths observed in practice with real metals. In fact, the shearing strength of a crystal of a pure metal may be only 2,000 to 3,000 pounds per square inch. Even ordinary commercial mild steel begins to yield at around 15,000 p.s.i. (0.01 MN/m^2), which is perhaps one-sixtieth of the theoretical value. Really strong steels might reach one-fourth of the theoretical value—say, about 250,000 p.s.i.—but at this point they are getting very brittle. Only under strictly controlled experimental conditions have some very strong glass fibers and some very strong whisker crystals of metal been observed to sustain shear stresses just below the theoretical values.

In an ideal or theoretical crystal, the planes of atoms are supposed to be stacked in a perfectly regular manner like the pages of a book. Perhaps, said Taylor, the arrangement of the planes of atoms in a real crystal is not actually as regular as we have thought. He went on to suggest that in most crystals there is, every now and then, a partial extra layer of atoms—much as if someone had slipped odd-shaped pieces of paper here and there between the pages of the book.

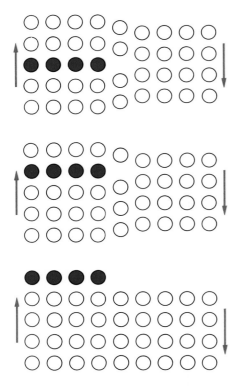

The shearing of a crystal by means of an edge dislocation. Top, edge dislocation. Middle, dislocation sheared one lattice spacing. Bottom, dislocation sheared out of crystal altogether.

Fundamental to Taylor's theory was the answer to the important question, what happens along the line where the extra layer of atoms comes to an end? The bonds of atoms in the neighboring layers are, in fact, strained in shear through an angle that is almost equal to the theoretical breaking strain. In other words, locally, the crystal is already practically broken in shear along the line of the dislocation. By applying quite a small additional shearing stress to the crystal as a whole, these bonds can be broken. But then we find that we have merely moved the dislocation one atomic spacing further on, while the broken bonds have re-formed behind it. The dislocation is thus a *mobile line defect*. By continuing to apply the shearing stress, we can frequently drive the arrangement right across the crystal until it emerges as a "step" on the surface.

Dislocations are abundant in practically all crystals, both metallic and nonmetallic. The nature of the metal-to-metal bond is such that after it has been broken it is comparatively easy to re-form. Thus dislocations in most metals tend to be very mobile. With most nonmetals, although the interatomic bonds are often stronger than metallic ones, they are usually not so easily re-formed. Consequently dislocations in nonmetals tend, with a few exceptions, to be much less mobile. As we shall see, dislocation mechanisms play a large part in the *growth* of nonmetallic crystals, but they do not, as a rule, influence their mechanical behavior very much.

Although I knew Sir Geoffrey Taylor fairly well, I am afraid that I spent too much of my time with him in discussing sailing and I never got around to asking how he had developed his ideas on dislocations. The concept of the dislocation is at the root of the understanding not only of the mechanical behavior of metals, but of many other important phenomena as well.

DISLOCATION SOURCES

Originally, Taylor seems to have supposed that the ductile behavior of metals was due entirely to the presence of edge dislocations that had formed by chance when the metal solidified. For instance, if two neighboring crystals grow with their atomic planes forming a small angle with each other, the two crystals will form a *small-angle boundary* consisting of a line of edge dislocations that may afterwards drift away. However, subsequent work convinced Taylor that these original chance-formed dislocations are not generally numerous enough to account for all the observed phenomena. In fact, new dislocations can be created in large numbers in a ductile metal at room temperature when the circumstances are favorable.

Fortunately for engineers, new dislocations can arise at severe concentration of shearing stress, such as John Cook found to exist on either side of the tip of a crack. Thus at critical or dangerous points in a metal structure, many hundreds of dislocations are liable to be born just where they are most needed. By this mechanism the stress concentration is generally relieved to considerable extent. New dislocations can also be generated in very large numbers by means of *breeding* mechanisms.

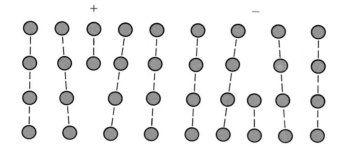

The sign of a dislocation. A dislocation can be either positive or negative depending on its location. Dislocations having opposite signs tend to attract each other and may join up, thus reducing the strain energy in a crystal. Dislocations having like signs repel each other, and if they are forced by other dislocations to approach each other, tend to increase the strain energy in a crystal.

We have defined a dislocation as a line of mechanical strain. Thus crystals contain strain energy—in fact, quite a lot of it. Because dislocations are able to "face" in either of two directions, they may be considered as having a sign: plus or minus.

If two dislocations opposite in sign approach one another, the strain energy in the system will be diminished. For this reason dislocations that are opposite in sign tend to attract one another, and if the crystal geometry is favorable, they may join up. Conversely, dislocations with the same sign tend to repel one another. In the busy dislocation commerce that goes on inside a metal under stress, these attractions and repulsions are very important. Among other reasons—as the British physicist Sir Charles Frank of Bristol University pointed out—they can cause dislocations to breed, sometimes very prolifically indeed.

One such breeding mechanism is the *Frank-Read source.* Suppose that a short dislocation is "pinned" at each end, either by two neighboring dislocations of like sign which happen to traverse the crystal in a different plane, or else by two fine particles of some impurity. Under an appropriate shear stress the dislocation line will be bent between the pinning points and, eventually, will bow out behind them. But the parts of the curve behind these pinning points will now, effectively, have opposite signs. So the two bowed arms of the dislocation will attract each other and may even join up in such a way that a new dislocation line is born. In fact, the process can be repeated indefinitely. Although the Frank-Read source might appear to be a highly improbable phenomenon, it has been observed in the

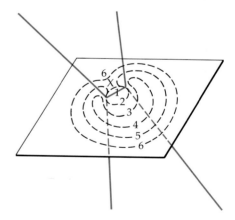

The Frank-Read source for generating new dislocations in a crystal. Two dislocation lines intersect the plane; they are joined by the red source dislocation line. Under a shear stress on the plane, this line tries to glide. Since, however, it is held back at its nodes by the immobile dislocation lines, it passes through a sequence of configurations (numbers 1 to 6).

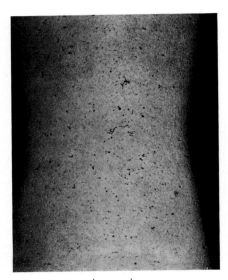

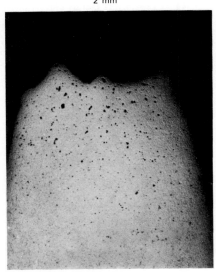

The progressive necking down of a copper specimen subjected to increasing tensile stress. Note the tendency for holes to appear in the highly strained region.

electron microscope and it is known to be very common indeed in ductile metals. One such source typically gives birth to about 500 new dislocations.

TOUGHNESS AND HARDNESS IN METALS

The strain energy of a dislocation in a typical metal can be calculated and is around 5.5×10^{-9} joules per meter. The ability of metal crystals to generate large numbers of dislocations when they are stressed enables the material to be shaped in useful ways and to yield locally to diminish severe concentrations of stress. It also means that to break a ductile metal requires a great deal of energy. In other words, the work of fracture W can be high.

In special circumstances, interactions between dislocations can act as a source of many new dislocations and thus, effectively, making the material softer. Much more often, as they get more numerous, dislocations get themselves into complicated tangles and tend to repel each other. In other words, the metal hardens. When this happens, although the material does get harder and stronger, it also gets more brittle because dislocation movement is now more restricted.

Although metals usually break in tension more or less on one plane, the resistance of a metal to fracture is definitely a three-dimensional phenomenon: it involves the creation of dislocations in regions that are very remote, in molecular terms, from the actual fracture surfaces. These dislocations need energy to create them. Most of the high work of fracture, W, of ductile metals comes from the creation of these myriad dislocations.

A typical reasonably ductile metal will "neck"—grow thinner in the fracture zone—before fracture, because of the shearing caused by the millions of newly made dislocations (see the figure on page 106). The number of dislocations created, and therefore the magnitude of the fracture energy, will be roughly proportional to the volume of the metal that is distorted on either side of the fracture plane. This volume will depend partly on the characteristics of the metal or alloy used and partly on the thickness of the specimen.

With plates or sheets of similar metals, the length of the "necked" zone will be more or less proportional to the thickness of the sheet. The volume that has been distorted will therefore vary approximately as the *square* of the thickness of the sheet, and so, in all likelihood, will the work of fracture. This is why thin aluminum foil tears easily even though it is made from soft metal, whereas the thicker aluminum sheets used for covering aircraft, for instance, are very much tougher. For metal plates more than about a centimeter thick, however, the necking zone does not get much longer as the thickness increases, so there is a limit to the improvement in toughness obtainable by thickening.

It must be reassuring to materials scientists that the works of fracture of common metals calculated from modern dislocation theory agree remarkably closely with the experimental values measured by traditional engineering methods. As we shall see, it is possible to count the number of dislocations present on the sectioned surface of a crystal under a microscope. For a fully cold-worked metal, this number is found to be in the region of 10^{16} dislocations per square meter. If the depth of the necking—the width of the plastic distortion on either side of the fracture surface—is, say, 1 centimeter (a typical value for a reasonably thick specimen of metal), then 1 square meter of fracture surface will contain about 10^{14} meters of dislocation lines. Because the calculated energy of a dislocation is somewhere around 5.5×10^{-9} joules per meter, the calculated work of fracture would be about 5.5×10^{5} joules per square meter. The experimental values for mild steel vary between 10^{5} and 10^{6} joules per square meter: a remarkable example of agreement between theory and practice.

As we said earlier, if we make any extensive use of the dislocation mechanism to shape or distort the metal at room temperature, we are making inroads upon the material's safety factor or reserve of toughness. Cold working (e.g., hammering) is the traditional method of hardening the cutting edges of tools and weapons, but if the load-carrying parts of a metal structure are expected to remain tough after they have been shaped to any considerable extent by cold working, they have to be *annealed*. The metal is heated to some temperature that is below its melting point but well above room temperature and then cooled rather slowly. If this is properly done, the metal will recrystallize and, in doing so, "erase" most of the many millions of dislocations that have been generated by the cold work. The dislocation density will be reduced from about 10^{16} per square meter to around 10^{10}, which may be regarded as a state of innocence. From this annealed or softened condition, the mechanisms needed for the generation of more dislocations through work hardening can begin again.

The changes in shearing strength brought about by work hardening and annealing have played a major part in history. The history books record, for example, that the Greeks won the battle of Marathon in 490 B.C. largely because the Persian arrows failed to penetrate the Greek armor. History books also record that the English won the battle of Crécy in 1346 and the battle of Agincourt in 1415 at least partly because the arrows of the English bowmen penetrated the armor of the French knights. Although most of the history books go to considerable lengths in describing the political and social consequences of these victories, I do not think that any of them have speculated on the technical causes.

The author's colleague Dr. Henry Blyth has done some research on the technological side of the battle of Marathon. Blyth used modern apparatus and techniques to examine many specimens of Greek armor. He found that the Greek armor, which was made of bronze, had been repeatedly annealed and was used in a very soft condition. This meant, of course, that the work of fracture was maximized. The Persian arrows therefore caused extensive denting when they hit but this distortion absorbed most of the kinetic energy of the arrows, which if they pierced the armor at all seldom penetrated far enough to inflict a dangerous wound. In fact, only 192 Greeks were killed at Marathon, whereas the Persians lost about 6,300.

The arrows from medieval English longbows are known to have had a higher kinetic energy than the ancient Persian arrows whose effects were examined by Henry Blyth. This was partly because the English longbowmen strung their bows with flax cords, whereas the Persians used animal sinew or tendon. Flax has a much higher Young's modulus—is much stiffer—than tendon. A recent computer analysis of bow design at Reading University shows that it is very beneficial to use a stiff string. Moreover, not only did the English arrows have a greater kinetic energy than the Persian ones, but the armor of the French knights of the fourteenth century was made from iron containing a fair amount of carbon, so it was almost certainly less tough and ductile than ancient Greek bronze. In any case, the expensive and prestigious armor of the French knights and men-at-arms was repeatedly penetrated. At Crécy an English army of 36,800 men defeated a French army of about 130,000. The English losses were slight, but about 30,000 Frenchmen were killed, including a great many knights and nobles.

Throughout long periods of history the balance between the efficacy of missile weapons such as arrows and the protection afforded by armor was a narrow one. The thickness of the armor that could be worn was limited by its weight, especially since military campaigns were generally conducted in summer, even in Mediterranean countries. Although the ability or the failure of missiles to penetrate armor has been enormously important in military and political terms, only quite recently have we begun to understand the underlying technical and scientific mechanisms involved—notably those of dislocation behavior.

SCREW DISLOCATIONS AND CRYSTAL GROWTH

When a substance crystallizes from its vapor or from solution, energy changes are involved. Whether or not crystallization takes place depends not only on the degree of supersaturation—that is, on how badly the atoms or molecules "want" to come out of the vapor or solution due to

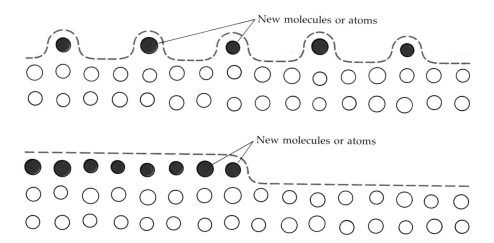

New molecules or atoms

New molecules or atoms

Step mechanism. Any change of state—such as from vapor to solid—must in general involve energy changes and energy barriers (e.g., latent heat). One kind of energy barrier is concerned with the necessary formation of new surfaces, as in crystallization. If molecules are deposited on a preexisting surface at random (top) there must be an increase, if only transiently, in surface area—and therefore in surface energy. This clearly puts a barrier in the way of the change of state. This energy barrier is largely evaded by the step mechanism (bottom).

overcrowding—but also on the nature of the substrate on which they are deposited.

The energy incentive or the degree of supersaturation required for atoms or molecules to be deposited on a uniform plane surface of a crystal is usually calculated to be quite high. Unfortunately for a simple theory, many crystals are observed to grow rapidly and healthily at supersaturations well below the calculated minimums.

In 1948 Professor Frank at Bristol postulated the *screw dislocation* as a mechanism that would enable crystals to grow more easily and rapidly. Although a considerable energy incentive is needed to make a solitary atom forego random motion in a vapor or solution to sit in isolation upon a wide uniform plane of similar atoms, the process becomes much easier if there is a preexisting molecular "step" on the crystal surface. This molecule provides a comfortable and welcoming nook into which a newly arrived molecule can settle at a much lower price in energy.

The step mechanism had been observed in operation by the British metallurgists C. W. Bunn and H. Emmett around 1949. The question it raised was, what happened to the step when it reached the edge of the crystal? In a crystal constructed of regular flat planes, one would expect the step and the growth process to halt upon reaching the boundary of the crystal. How could the step be regenerated for a continuous growth mechanism?

Frank suggested that in many crystals the molecules are not organized in parallel, finite layers like the brickwork of a building, but rather in a helix, an arrangement like a flat spiral staircase. In a helical structure the

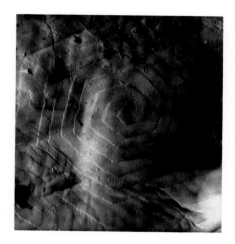

Screw dislocations not only affect the mechanical properties of a crystal, they also facilitate crystal growth by providing a continuous growing edge.

METALS AND DISLOCATIONS

Dendritic crystals on the surface of an ingot of antimony.

step or growing edge would never have to come to a halt, so the crystal could grow continuously in the form of a screw. This brilliant idea turned out to be true. Screw dislocations were observed by John Forty and other colleagues of Frank around 1951.

Although it is perfectly possible for crystals to grow without using the screw dislocation mechanism, this mode of growth is in fact a very common one both in metals and in nonmetals. Most often a crystalline material grows from many separate screw dislocations simultaneously into the familiar irregular polycrystalline arrangement of metals and many minerals. Occasionally, a growing screw dislocation may become detached from its surroundings and grow independently into a thin, smooth whisker fiber that is millimeters or even centimeters long (Chapter 4). These whiskers are often very thin, but sometimes the original filament is thickened by additional growth layers that spread along the crystal until they reach the tip.

Thin whiskers of most substances tend to be very strong indeed and may approach the theoretical tensile strength of the material. The high strength arises partly because the surface is generally very smooth and partly because the geometry of the whisker may not allow dislocation sources to operate. As a whisker crystal gets thicker, its strength usually falls off.

Although the dislocation that runs along the middle of a whisker is a simple screw dislocation, many of the dislocations in larger crystals are likely to be more complicated, because there is no absolute distinction between a screw and an edge dislocation. In fact, an edge dislocation can be something in between, so that it has both an edge component and a screw component. Dislocation lines can wander about a crystal and zigzag back and forth in all three dimensions. The laws of the movement, attraction, and repulsion of these dislocations tend to be a bit involved: what goes on within a simple-seeming little crystal is more elaborate than three-dimensional chess.

Not only do dislocations have very important effects upon the strength and the growth of crystals, but they also have significant chemical effects. Because the interatomic or intermolecular bonds along the line of a dislocation are highly strained—especially in the center of a screw dislocation—they are particularly vulnerable to chemical attack by acids, by corrosion processes, and by other reagents. Moreover, an emergent dislocation, especially a screw, can promote chemical reactions that do not attack the material of the substrate—that is, the crystal containing the dislocation. Thus dislocations play an important role in chemical catalysis.

It is interesting to reflect on the number of ways in which Nature uses helical geometries which, in engineering, are largely confined to nuts, bolts, and ropes. The analogy between the screw dislocation and the famous double helix of Crick, Watson, et al., in genetics is obvious. As we shall see in Chapter 6, a helical morphology plays a very important part in the remarkable toughness of trees and other plants. The keratin structure of animal hair is helical, and so are a number of other molecular and larger structures that are common in biology. Engineers should probably think about helices more than they do, instead of automatically reverting to orthogonal solutions.

THE OBSERVATION OF DISLOCATIONS

Dislocations are a perfectly real physical phenomenon that can be observed in the laboratory in various ways. One technique is chemical etching. Because the bonds near a dislocation line are considerably strained, they are more easily broken by chemical or physical means than the bonds in the unstrained parts of the crystal. If the surface of the crystal is treated with an etching agent such as a weak acid solution, the points at which the dislocations emerge will be attacked more severely by the acid than the unstrained parts of the surface, and will thus often show up as a series of pits that are easily seen with an optical microscope.

Another fairly popular technique for studying the movement of dislocations is to split a crystal in two. Any dislocations that existed in the crystal before the experiment will naturally give rise to matching patterns on the cleavage surfaces. However, if half of the crystal is kept as a control while the other half is distorted or otherwise experimented on, any experimentally caused dislocation movement or newly born dislocations will show up when the two surfaces are etched and compared.

More detailed pictures can be obtained with the electron microscope. Some of the earliest photographs of dislocation lines in thin metal films were taken by Dr. Peter Hirsch working at the Cavendish Laboratory in Cambridge, England, in the early 1950s. The dislocations show up as a network of thin black lines on a white or gray background. Because the metal film was heated by the electron beam and thus subjected to thermal stresses, the dislocations could be seen to move. Hirsch was able to make movies of this behavior. The viewer gets the impression of a lot of scurrying mice: the movement is rapid, complicated, and very extensive.

Hirsch's pictures show the dislocations as dark, slightly blurred lines, and do not give any indication of their detailed molecular structure. In a normal metal crystal, the lattice spacing is typically around 2 angstroms (abbreviated Å; 1 Å $= 10^{-10}$ m). In 1955, when Jim Menter began to work on dislocations a few miles down the road from Hirsch, the best resolution of the electron microscope was around 10 Å. There seemed to be no hope of observing the structure of dislocations on an atomic or molecular scale.

An edge dislocation in a crystal of platinum phthalocyanine appears in the left third of the photograph, slightly above the middle.

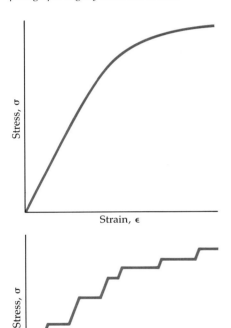

Above, a normal macroscopic load-extension curve for a ductile substance. Below, a load-extension curve for the same material tested on a fine scale in the Marsh machine. Plastic extension is broken down into steps, each of which corresponds to the operation of a dislocation source.

However, Menter made use of a substance called platinum phthalocyanine. The phthalocyanines are a family of organic compounds that have large, flat molecules—shaped rather like playing cards—about 12 Å across. In the middle of the molecule there is a hole that can accommodate a metal atom—in this case, platinum. Platinum, of course, is much denser than the carbon, oxygen, and nitrogen atoms of the organic part of the molecule surrounding it. Platinum phthalocyanine crystallizes fairly easily with a molecular spacing of 12 Å. The crystal therefore contains regular, parallel lines of platinum atoms, also 12 Å apart. By adjusting the electron microscope to its best resolution, Menter was able to resolve these lines of platinum atoms. The micrographs showed enormous numbers of slightly blurred dark gray lines, all perfectly parallel. Only after many such pictures had been taken was an edge dislocation found. It looked exactly like it should have: one dark fuzzy stripe came to an end, and the neighboring stripes closed in around it. Jim Menter was able to send this photograph to Sir Geoffrey Taylor—the original dislocation theorist—in time for his seventieth birthday.

A year or two later, David Marsh, also working at Hinxton Hall, was able to detect the operation of dislocation sources mechanically. As we mentioned in Chapter 4, Marsh had developed a microtesting machine of extreme sensitivity. If a normal-sized specimen of a ductile metal is pulled in an ordinary engineering testing machine, the resulting load-extension diagram is a smooth curve as the ductile behavior becomes more and more pronounced. When a very small specimen such as a thin whisker is tested in the Marsh machine, the resulting load-extension diagram is a straight line, because there is no room for dislocation sources to operate. The material has to obey Hooke's law.

With a slightly thicker whisker, in which there is just room for a limited number of sources to operate, the load-extension diagram is neither a smooth curve nor a straight line. It is a series of steps, each step corresponding to the action of a single dislocation source. For the operator of the Marsh machine during such a test, the speed with which each yield or step occurs is most impressive. Dislocation sources operate very rapidly indeed.

FATIGUE FAILURE

Although dislocation mechanisms are responsible for many of the mechanical virtues of metals, they are also responsible for some of what Samuel Butler, in "Hudibras," called their "dog tricks." As we have seen, it is possible to work-harden a metal, for instance by cold work, to such an extent that it becomes dangerously brittle. This is an error or sin to which modern manufacturers are prone, because annealing metal components after they have been pressed into shape adds to production costs. We have also referred to another dog trick of dislocations: their tendency when present on the surface of metals to expedite corrosion.

More subtle, and usually more dangerous, is the liability of metals to failure in *fatigue*—a weakened state brought about by repeated stressing. A structure that has been in service for several months or years and has come to be considered absolutely safe may fracture quite suddenly. Because such failures generally occur without any warning, they are very dangerous and have been responsible for much loss of life.

As Griffith pointed out, for a metal to be safe in tension it must have a high work of fracture. The total amount of energy required to break a ductile metal is considerable. If a new structure is to be broken immediately, all this energy has to be produced more or less at once. Such situations are the subject of classical fracture mechanics.

Unfortunately, it is also possible to pay the energy price of fracture not with a single lump sum payment, but by installments. If they are numerous enough, each installment of energy can be very small. Fatigue thus arises when a metal structure is subjected to many millions of applications or reversals of a load that would be perfectly safe if it were applied only a few times.

Although fatiguelike processes can and do occur in materials such as wood, they are much less common and less dangerous in nonmetals. Consequently, fatigue failures were not noticed by engineers before the use of rotating metal machinery became common in the middle of the nineteenth century. The classic work on fatigue was carried out by August Wöhler (1819–1914), a German railroad engineer.

Many of the parts of a railroad train—especially the axles of the cars—are subjected to fluctuating loads. In Wöhler's time these axles would break quite suddenly after fairly long periods of service, sometimes causing serious accidents. Wöhler investigated the phenomenon of fatigue failure in steels in a practical, quantitative way. In doing so he provided a valuable service to engineers and saved many lives.

In his numerous tests he made use of a metal specimen in the form of a rotating shaft or cantilever. On its projecting end he hung a weight, attached and supported by a ball bearing. By fitting a revolution counter to the shaft and making use of simple beam theory, he was able to determine the number of reversals of loading that the material could withstand at any given stress before failure. When Wöhler came to plot his results in the form an S-N diagram (S here stands for strength, N for number of reversals), he found that for iron and steel there is a *fatigue limit*—a level of stress below which the metal will not fail by fatigue, no matter how many million times the load is reversed.

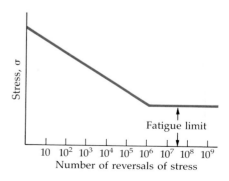

Typical fatigue curve for iron or steel.

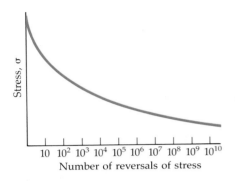

Stress, σ

10 10^2 10^3 10^4 10^5 10^6 10^7 10^8 10^9 10^{10}
Number of reversals of stress

Typical behavior of a nonferrous metal in fatigue.

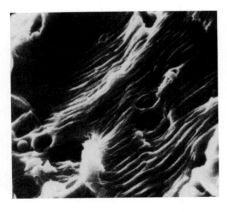

Ductile striations in mild steel. The general direction of crack growth is from top to bottom.

The fatigue limit is, of course, a valuable property in steel, and designers take advantage of it. Unfortunately, when aluminum alloys were developed a few years after Wöhler's time, it was found that nonferrous—non-iron-containing—metals do not generally have a definite fatigue limit. Instead of flattening sharply somewhere between 1 and 10 million reversals, the S-N curve usually trails off gradually. This characteristic is one good reason for the use of large factors of safety in airplanes.

The fatigue process begins on the surface of a metal, generally—but not always—at a preexisting stress concentration. A very fine crack spreads very slowly from the surface down into the metal. Once this fatigue crack reaches a length that corresponds to that of a normal, "static" Griffith crack, it propagates suddenly and explosively right across the component. Because in the early stages fatigue cracks are very thin and often grow in hidden parts of the structure, they may be extremely difficult to detect at routine inspections. After failure has occured, however, fatigue cracks are normally very easy to recognize because the broken surfaces tend to have a characteristic banded appearance.

In recent years the fatigue process has been the subject of much investigation in the light of modern dislocation theory. The mechanism is apt to be complicated and it can vary greatly in detail. One of the points to emerge is that a fatigue crack may be initiated from a very minor concentration of stress existing on an almost smooth surface. Such a location will be vulnerable if a shear stress acts at roughly 45° to the surface and relatively few edge dislocations are waiting in the right places to absorb the stress. The Cottrell-Hull mechanism explains how mobile dislocations can be transformed into actual cracks in a crystal as a result of small stresses.

Fatigue cracks in their later stages can sometimes be detected by acoustic means. Train passengers of an older generation may remember being awakened during the night in our sleeping cars when the train stopped at some station and railwaymen went along the track hitting the car wheels with hammers. If they "rang true," all was well. Going back further into the past, metal fatigue has left a well-known mark on a symbol of American history. The fatigue crack in the bronze Liberty Bell in Philadelphia apparently took only 69 years to develop.

Although it is possible to produce fatigue failures in wood in the laboratory, fatigue does not generally seem to present Nature with any serious problems. Trees, for instance, are buffeted to and fro by the wind, and in the course of their long lives (which sometimes extend for thousands of years) they suffer many millions of reversals of load, but they do not seem to be much the worse for it.

Curiously, the wood used in musical instruments is liable to a form of fatigue. For instance, after prolonged use a violin becomes "played out"

FATIGUE FAILURE IN LITERATURE

Nowadays the prevention of fatigue failure by good design and the investigation of possible fatigue failure in accidents are important preoccupations of professional engineers. In human terms the dramatic possibilities of fatigue failures have been rather slow to appeal to the literary profession. Probably the first nontechnical story about fatigue, published in 1895, was Kipling's "Bread upon the Waters," about a fatigue crack in a ship's tailshaft. Kipling has become an unfashionable writer, but his understanding of the human and dramatic implications of engineering is almost unequalled, although it is true that Kipling sometimes slips up on technicalities.

Fatigue failure as a literary theme does not seem to have been revived until a novel by the sometime aircraft engineer—who had worked at Farnborough—Nevil Shute's *No Highway*, was published in 1948. A dramatic story about fatigue failure in a transatlantic airliner, *No Highway* was a success as a novel and was made into a movie which was also successful, partly perhaps as a result of the publicity associated with the accidents to the Comet airliners which happened not very long afterwards. In various forms *No Highway* has been repeatedly broadcast both on radio and television down to the present time. It must have done quite a lot to educate the public about one of the basic facts of life and death in engineering.

because of irreversible changes in the wood of its body, which then ceases to resonate properly. This is why ancient and valuable violins should not be played casually.

METALLURGY

Notionally, the subject of metallurgy falls into two divisions or categories: *extraction metallurgy*, which focuses on the separation of the metal from the other elements with which it is mixed or combined in its ores, and *physical metallurgy*, which is chiefly concerned with ensuring that the metal possesses the right combination of mechanical and other properties for its intended use. Hypothetically, extraction metallurgy is chiefly a chemical problem, whereas physically metallurgy is based on the control of dislocations. In practice, the division of labor is seldom so simple. Commercial extraction processes frequently do not yield the metal in a pure state, so the two concepts or divisions of the subject have to be considered together. The whole business is prone to Byzantine complications that are further exacerbated by traditional and commercial considerations. A vast number of books have been written on the subject, some of them more readable than others. Lest we get lost in a wilderness, we shall deal here only briefly with the problems of traditional metallurgy, for we are concerned mainly with the mechanisms of strength, which do not differ much, in principle, in the various metals of technology.

The first metal to be used by humans for mechanical purposes, such as tools and weapons, was almost certainly copper. Copper sometimes occurs *native*—that is, in the uncombined metallic state. When he was exploring Greenland in 1894, Robert E. Peary found that the local Eskimos were in the habit of breaking off pieces from a very large copper meteorite and using them to make tools. There was still about 37 tons of it left when Peary arrived. It was eventually shipped to the American Museum of Natural History in New York.

Even larger copper meteorites have fallen in other places, but as a source of supply meteoric copper is limited and unreliable. Although it does occur native in small quantities elsewhere—mainly in the Middle East—a far more important source, both in the ancient world and today, has been the mineral copper pyrites ($CuFeS_2$). Much of the copper in antiquity came from Cyprus, and it is often said that the island took its name from the Latin word for copper. In fact, the metal was called *copper* because it came from Cyprus, which was actually one of the names of the goddess Aphrodite.

Somewhere around 3500 B.C., possibly in Cyprus, metallic copper was produced from copper pyrites by reducing the ore in wood or charcoal fires. Pure copper melts at 1083°C. It is somewhat viscous as a liquid and not particularly easy to cast. Early on it was discovered, no doubt accidentally, that the addition of a few percent of tin both lowered the melting temperature and made the metal easier to cast. More importantly, better combinations of strength, toughness, and hardness could be achieved.

Copper-tin alloys are, of course, traditional bronze. In recent times the appellation has been extended to a variety of other alloys of copper and of aluminum. The supply of tin-copper bronze for weapons and armor became militarily and politically important in antiquity. Copper was reasonably plentiful in ancient times in the Middle East, but tin occured only in small quantities in Cyprus and in what is now Turkey, and it soon became scarce. As a result both the Greeks and the Phoenicians made long and dangerous voyages through the Straits of Gibraltar and across the Bay of Biscay to Britain—the Cassiterides or "Tin Islands." The Cornish tin industry prospered accordingly, and in various forms these mines had been worked continuously until the 1980s. After some 3,000 years they are no longer profitable, and they are being shut down and abandoned amid an outcry from antiquarians, trade unionists, and the general public.

The copper industry in the United States is , to some extent, a spin-off from the gold-rushes of the middle of the nineteenth century. Sometimes the prospectors were disappointed to find extensive deposits of copper ore instead of gold. The less astute entrepreneurs sold out cheaply to shrewder men like Marcus Daly, who founded, among other things, the Anaconda Company.

Although for tools, armor, and cutting weapons, cheaper iron and steel largely ousted bronze by the middle ages, it remained in use for church bells, partly because it was easier to make very large castings in bronze and partly because the medieval church was rich enough to afford such things. From church bells to cannon is a natural step, technologically; some of the best ships' cannon were cast from Admiralty bronze or gunmetal. Bronze cannon were always better and more reliable than cast iron ones, which were liable to burst when fired. Nowadays real bronzes are used mainly for ships' propellers, for statues and monuments—and for church bells.

Brass, a copper-zinc alloy containing up to 50 percent of zinc, is much cheaper than bronze, but its mechanical properties are generally not so impressive: brass is apt to be rather brittle and liable to fatigue failure. However, its corrosion resistance is fairly good and it can be plated with chromium or nickel to make it attractive for many purposes. In the modern era both copper and brass are used less for their mechanical properties than for their combination of corrosion resistance and high electrical conductivity.

Photomicrograph of brass.

HISTORICAL IRON-MAKING

The nomenclature of iron and steel is apt to be confusing. *Iron* by itself usually means iron in a relatively pure state. *Pig iron* and *cast iron*, on the other hand, are forms of iron containing nearly as much carbon as they will hold, usually about 4 percent. *Wrought iron,* which is now mostly a historical material, is a fairly pure iron containing glassy inclusions. *Steel* usually means iron with a small, controlled amount of carbon in it, generally less than 1 percent. *Alloy steel* is a general term for steels alloyed with elements other than carbon. Alloy steels often have properties superior to those of carbon steels, but they are nearly always much more expensive.

In steels, therefore, the content of carbon or other alloying elements is under close control. The movement of dislocations, and thus the mechanical behavior, of iron is profoundly influenced by the addition of other elements. Keep in mind that the amount of another element added is always quoted as a percentage by weight, which may be deceptive. The carbon atom has roughly one-fifth of the weight of an iron atom, so 4 percent of carbon by weight is about 20 percent by volume or by number of atoms.

Although we talk about *cold iron*, much of the technology of iron and steel is a technology of high temperatures. Copper melts at 1083°C, as we have said and bronze usually between 900 and 1000°C—temperatures that are fairly easy to reach with wood or charcoal fires. Pure iron melts at 1535°C, which was well beyond the capability of ancient furnaces. However, when 4.0 to 4.5 percent of carbon is diffused into the metal, its melting point may be reduced to about 1150°C, a temperature that can just be attained by a *blown* charcoal fire—that is, a fire encouraged by blowing on it.

The most common form of iron ore is hematite, Fe_2O_3, which takes its name from its blood-red color. (It is used as a red pigment in paints.) The industrial preparation of pig iron begins with the heating of hematite with an excess of carbon in the form of charcoal or coke. The following two reactions take place:

$$3Fe_2O_3 + 11C \longrightarrow 2Fe_3C + 9CO$$
$$Fe_2O_3 + 3C \longrightarrow 2Fe + 3CO$$

The carbon monoxide goes off as a gas and we are left with a mixture of Fe_3C—cementite—and elemental iron. Cementite and elemental iron are mutually soluble and produce a substance that contains about 4 percent carbon by weight and that melts, as we have said, at about 1150°C. This, basically, is pig iron. Iron ores also contain many mineral impurities, largely oxides of other metals. These often have high melting temperatures. To eliminate these impurities a combining agent or flux is added, generally lime or limestone. The flux reacts with the impurities to form a glass that has a fairly low melting temperature and a considerably lower specific gravity than that of the molten iron. As a result, most of the nonferrous impurities float to the top of the furnace in the form of *slag*, which is relatively easy to remove.

The molten mixture of cementite and iron is run off from the bottom of the furnace and cast, either into pigs—rough bars—for conversion into wrought iron or steel, or else directly into shaped molds for military, technological, or artistic purposes. The history of cast iron is a long one. The

Roman author Pausanius, writing in the second century A.D., notes that "Theodorus of Samos was the first to discover how to pour or melt iron and make statues of it." Various modern critics have doubted whether ancient furnaces could have reached the 1150°C needed to melt pig iron. However, a few years ago some engineering students of the author's, working in conjunction with classics scholars, constructed what they hoped was a reasonably accurate replica of an ancient iron furnace. Burning charcoal, this furnace was able to achieve temperatures around 1300°C. So it seems quite likely that Pausanius was right.

The early ironmaking furnaces were quite small; they burnt charcoal or wood, and they were blown by hand, using either a fan or a bellows. By the fifteenth century the furnaces in Western Europe were getting larger, and waterwheels began to be used to drive the bellows.

Nowadays simple pig iron is generally considered to be too brittle for most structural purposes. Cast iron components, such as the cylinder blocks of automobiles, are made from pig iron whose composition has been modified in various sophisticated ways so as to make it considerably tougher and stronger. In fact, most of the pig iron made does not end up in castings at all, but goes through further processes to convert it to steel by the removal of most of the carbon and, frequently, by the addition of other metallic elements.

Although historically cast iron was used, for lack of a better material, to make structures like cannon and arched bridges, it was always too weak in tension and much too brittle for use in most tools and weapons. As we have said, pig iron directly from the furnace contains a high proportion of cementite, Fe_3C. The dislocations in cementite are immobile at normal temperatures. Furthermore, crude cast iron is apt to contain inclusions such as thin plates of graphite, which act as built-in stress concentrations. For most purposes it is necessary drastically to reduce the carbon content of the iron and also to get rid of as many of the inclusions and impurities as possible.

To this end the early smiths heated the crude iron to a temperature in the region of 800 to 900°C, and then beat it out on an anvil. The hot metal was hammered into a slab or plate whose surface became covered with a layer of iron oxide. When the smith folded the plate over and continued to hammer it, the oxide was forced into intimate contact with the hot metal and some of the cementite was reduced to metallic iron:

$$Fe_3C + FeO \longrightarrow 4Fe + CO$$

For a single such operation the amount of carbon removed was quite small and the effect was a local one. Therefore the process had to be repeated many times; in the making of high-quality swords, perhaps thousands of times. Ultimately nearly all of the carbon would be removed, leaving only strings and filaments of glassy slag which were, on the whole, beneficial.

Iron can be melted and poured to make statues and decorative objects. Eugène Atget photographed this seventeenth-century doorknocker at 5, rue de Mail, Paris, in 1908.

THE DARBY FAMILY: THREE GENERATIONS OF IRONMASTERS

Curiously, the use of charcoal in ironmaking lingered, locally, in England down to quite modern times. For instance there was an ironworks at Backbarrow, near Lake Windermere in Westmoreland, which used charcoal until 1920. The furnace was blown by water power. Much of the charcoal which was used was made on my grandfather's estate a few miles away. I remember, as a child, that the charcoal burners who worked in my grandfather's woods were frequently bitten by the snakes which emerged from the wood-piles when they were set on fire to make the charcoal.

Much of the early development and subsequent expansion of pig iron and cast iron production in England can be credited to three generations of the Darby family. The first Abraham Darby (1677–1717) set up an ironworks at Coalbrookdale in Shropshire, England, around 1709 and developed improved methods of making iron castings in sand molds. His son, the second Abraham Darby (1711–1763), introduced the use of coke as a fuel in place of charcoal—an enormously important innovation. Until then the iron industry in England had depended on the rapidly diminishing forests of southeast England. With the ability to use coke, a much cheaper substance, in blast furnaces the British iron and steel industry moved from the woods of the south to the coalfields of the industrial north. The change to coke came a good deal later in continental Europe, and this was a factor of some significance during the Napoleonic wars.

The third Abraham Darby (1750–1791), also of Coalbrookdale, was responsible for the first really large iron structure, the famous Iron Bridge which opened in 1779. The arch design was selected because such a structure is almost entirely in compression, and cast iron—

Abraham Darby's Ironbridge over the River Severn near Coalbrookdale, Shropshire. This arch bridge was built entirely of cast iron, with a clear span of just over. 100 feet, a total length of 196 feet, and a height of 50 feet. It was built in three months in 1791 and cost 6,000 pounds sterling.

particularly traditional cast iron—is weak and brittle in tension. However, in spite of this, cast iron was used (largely because bronze was expensive) for critical tension structures, such as cannon, for a great many years. As we have said, a good many cast–iron cannon burst in action. It is interesting that, though French naval architecture and ship-building throughout the eighteenth century was usually superior to that of the British, French naval cannon were generally less reliable. The bursting of cannon aboard the French warships was influential at the battle of Trafalgar in 1805.

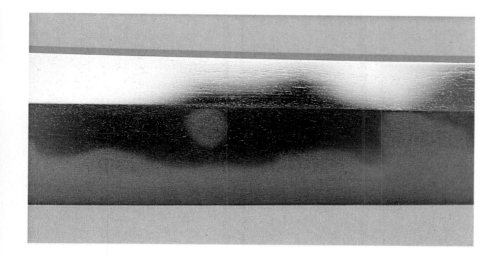

Wavy pattern on a Japanese sword resulted from folding and bending during reworking. The blade was polished to reveal the pattern of hard and soft metal.

This is why Japanese swords, for instance, often show a wavy pattern on the surface.

The result of repeatedly heating and working the metal was a fairly pure iron in which the dislocations could move freely. The material was tough but quite soft. Consequently the cutting edges of most tools and weapons had to be *carburized*; that is, carbon had to be put back into them locally. Historically this was done by packing the artifact round with carbon—and sometimes with other "magical" ingredients of doubtful efficacy—and reheating it. In the hands of a skillful smith carbon would diffuse into the iron to a depth of perhaps a millimeter, hardening the surface while leaving the interior in a tough condition. This process of *case hardening* is extensively used today, although the magical ingredients have changed.

The process of beating out iron by hand was, of course, very laborious, and throughout the middle ages iron tools, weapons, and armor were extremely expensive. Water-driven mechanical hammers invented in England during the fifteenth century reduced the cost of iron somewhat. The real breakthrough came with the invention of the *puddling* process by Henry Cort (1740–1800), whose patent dates from 1784. The puddling process made wrought iron relatively cheap and plentiful. In fact, the Industrial Revolution depended to a large extent on two things: the introduction of coke in blast furnaces by Abraham Darby and the invention of puddled wrought iron by Henry Cort.

In the puddling process the pig iron, instead of being laboriously beaten out, was melted into a pool or puddle. Into this puddle the workers, who were called puddlers, stirred iron oxide by means of a long tool called a rabble. When the oxide was stirred in, the bulk of the carbon in the pig iron was removed, as carbon monoxide, by the same chemical reaction that would have taken place in the hammering process, but much less slowly and laboriously. The iron could eventually be lumped into a ball weighing 100 pounds (about 45 kg) or so and removed from the furnace to be rolled into plates or rods. Although the productivity of the puddling process does not begin to compare with that of modern steelmaking, a skilled puddler could produce perhaps a ton of iron a day, representing a 10- to 20-fold increase over previous methods.

Puddled wrought iron was more expensive than modern mild steel and somewhat lower in tensile strength, but it was a very tough and reliable material, and one that was much more resistant to corrosion than most modern steels are. Wrought iron seems to last indefinitely. The steamship *Great Britain*, for instance, which was built in 1843, is in excellent condition

The steamship Great Britain, *built in 1843 (right). Though beached for many years in the Falkland Islands (left), the ship is still in excellent condition and is being salvaged.*

today, despite the fact that she lay neglected and unpainted on a beach in the Falkland Islands for many years. Modern ships and oil rigs, made from mild steel, often give serious trouble from corrosion within 10 years or so. The cost of scaling, painting, and other maintenance required by modern steel has to be set against its lower basic cost.

MODERN STEELMAKING

Puddled wrought iron was an excellent material—and, as we said, the "key" material of the Industrial Revolution—but it was always expensive. This was especially the case in the United States, where large structures, such as ships and railroad bridges, continued to be made mainly from timber until near the end of the nineteenth century. (Timber was not only cheaper and more widely available in North America, but the climatic conditions on that continent are such that outdoor wooden structures require less maintenance than they do in Western Europe. We shall see why this is so in Chapter 7). Even in England, the manufacture and use of modern steel on a large scale and at low prices is not really much more than 100 years old. The dominant position of the steel industries in most of the technologically developed countries today is a comparatively recent phenomenon.

The dry climate of North America—drier than the climate in Europe—allows the construction of large wooden structures, such as the covered bridges found across the contineut.

Bessemer converter.

Until the middle of the nineteenth century, such "steel" as was made was prepared in small clay crucibles in batches weighing up to about 90 pounds. Crucible steel was expensive, and it could not be considered as a feasible material for large structures. The development of a process capable of producing steel in tonnage quantities at low prices was largely the work of Sir Henry Bessemer (1813–1898). Bessemer was a prolific inventor with a considerable commercial flair. He had already made money out of inventing a "gold" paint and "lead" pencils (the "lead" was actually consolidated graphite). He became interested in steelmaking as a result of the publicity about the weakness of the cast iron guns used during the Crimean War (1854–1856).

Bessemer had the revolutionary idea of removing the excess carbon and other impurities from the liquid pig iron by blowing air through it. In the Bessemer process, about 30 tons of pig iron was run directly from the blast furnace into a large pear-shaped vessel, a *Bessemer converter,* in which air was forced up from the bottom through the molten iron. The "blown" air bubbles first oxidized the silicon and manganese impurities in the pig iron. These oxides formed a slag that floated to the top. After a few minutes the oxygen in the air began to react with the carbon, producing long, whitish-yellow flames. When the carbon had been eliminated, the flame dropped and the blast was turned off. These oxidizing processes produced enough heat to raise the temperature of the metal considerably, and a certain amount of scrap steel was added to cool the system and to prevent the ceramic material lining the converter from being damaged.

The original Bessemer process had one serious flaw: it was unable to cope with the sulfur impurities that are generally present in pig iron. Sulfur does not, as one might expect, oxidize harmlessly to SO_2, which would go off as a gas, but instead forms iron sulfide, FeS. When the molten steel solidifies, FeS separates out at the crystal boundaries, where it has a serious weakening effect. The cure for this problem is to add manganese, which forms manganese sulfide, a compound that passes into the slag and so is eliminated.

It happened that another British inventor, Robert Mushet (1811–1891), was developing a similar steelmaking process around the same time. The master patents Bessemer took out in 1855 contained no provision for dealing with the sulfur impurities. Mushet took out a series of patents for a very similar process in 1856. However, Mushet's inventions included the addition of *spiegeleisen,* a special form of iron containing manganese, to the molten metal. This practice solved the sulfur problem.

Bessemer's main patents had indisputable priority, but Mushet's patent on the addition of spiegeleisen should still have been quite valuable. Unfortunately, Mushet, by omitting to pay the stamp duty, allowed his critical patent to expire. Bessemer ruthlessly appropriated Mushet's idea without paying him any royalty. Bessemer ultimately made a huge fortune out of his steelmaking process. Toward the end of his life, he did pay Mushet, who was living in poverty, a pension of 300 pounds sterling a year, but Sir Henry Bessemer's reputation as an honest man had suffered considerably.

The eventual effect of the introduction of Bessemer steel was momentous, but not as rapid as one might have expected. Although the price of steel fell by a factor of at least 10 after the institution of the Bessemer process, about 30 years passed before mild steel superseded puddled wrought iron. Until about 1880 the British Admiralty refused to build warships in steel, partly because steel was considered, with some justification, mechanically unreliable, and partly because of its relatively high rate of corrosion in sea water. The first use of steel for a large and important structure was in the erection of the Forth railway bridge, built in Scotland in 1889. This bridge, which contained over 50,000 tons of steel, is still in place and working a century later. However, the amount of painting it requires to preserve it from rust has been a stock joke in Britain for at least three generations.

Bessemer steel made its first entry into the United States rather dramatically around 1862, in the shape of fast steamers built in British shipyards in order to run the Northern blockade into the Southern ports. This the ships did with considerable success, having a top speed of 20 to 22 knots because of the weight saving achieved in their hulls and machinery by the use of steel. The fastest ships of the more conservatively constructed Union navy

The railway bridge over the Firth of Forth, Scotland, built in 1889.

had a top speed of around 15 knots. The actual manufacture of Bessemer steel in the United States was somewhat delayed, no doubt by the effects of the war. The Bessemer process was introduced industrially in the United States by Alexander L. Holly between 1867 and 1870.

Another steelmaking process emerged around the same time as the Bessemer process and eventually superseded it for many years. The *open-hearth* process was originally developed by another prolific Victorian inventor, Sir William Siemens (1823–1883), and his brother Fredrick Siemens (1826–1904). In its original form it was really little more than a large-scale development of the puddling process, with the "puddle" of molten iron becoming more like a swimming pool. The furnace was heated, usually by gas, in an arrangement whereby the heat was recycled to reduce the fuel costs. In the Siemens device the outgoing hot gases were passed through a brickwork labyrinth in which they deposited much of their heat. At intervals the direction of gas flow in the system was reversed so that the heat deposited in the brickwork was used to preheat the incoming air. By means of this arrangement, furnace temperatures in the region of 1500°C could be achieved with a fairly moderate fuel consumption.

When the Siemens process was first put into operation in 1857 it was barely competitive with the Bessemer system, because in spite of the recycling arrangement the Siemens furnace needed a lot of fuel whereas the Bessemer furnace needed none. However, as the use of steel became more widespread, more and more scrap steel came on the market, until eventually around 50 percent of the total production found its way back to the steelworks as scrap. Scrap steel is not only a cheap raw material; it is also rusty, so its use reduces the consumption of iron ore in the converter.

In the Bessemer process, the oxidation of the carbon in the pig iron during the "blow" produces a comparatively small surplus of heat that is capable of melting perhaps a 5 percent addition of scrap steel. In its original form the Bessemer converter could not cope with anything like an intake of 50 percent of scrap. The use of scrap steel on a large scale in the Siemens process was the contribution of a Frenchman, Pierre Martin. Steel made in this way is therefore generally referred to as *Siemens-Martin* steel.

The use of large quantities of cheap scrap steel in the Siemens-Martin process more than offset the cost of the extra fuel that was required. Moreover, as compared with the Bessemer process, a considerably closer control over the composition of the final steel product could be exercised. In consequence, open-hearth furnaces tended to take over from the Bessemer system until, by about 1950, about 85 percent of the plain carbon steels in the world were open-hearth steels.

Recent years have seen the development of various furnaces that are cheaper and in some ways more convenient than the open-hearth system. There is a tendency to revert to "blown" furnaces that are somewhat similar in principle to the Bessemer converter, except that they are generally blown with oxygen rather than with air. As a result, the production of traditional open-hearth steel has fallen fairly dramatically. Another new development is to blow the open-hearth furnace with oxygen to speed up the process. Electric furnaces provide yet another alternative, particularly for the manufacture of smallish quantities of special high-grade steels.

THE VIRTUES AND LIMITATIONS OF IRON AND STEEL

Politicians and industrialists, ordinary people, and many engineers who ought to know better are prone to thinking of steel as a material with a special status. It has been held up as a symbol of progress, with its wider and wider use being promoted as desirable and inevitable. At the root of this belief is the assumption that there can be a single universal material that is best suited to load-bearing devices of practically any kind. Because steel can be made stronger than most other materials, the reasoning goes, it must be better. The problem of sustaining a load efficiently and economically is far more complicated than that, as we have seen throughout this book. Actually, we need only consider the immense variety of solutions to structural problems that have evolved in biology.

The very characteristics that have rendered iron and steel so valuable in certain specialized areas of technology have made them unsuitable for use in many wider contexts. Steel is a dense material with a specific gravity of 7.8, as opposed to 1.0 to 1.5 for most biological materials. It is able to pack a great deal of strength and stiffness into quite a small bulk. This is clearly a valuable property in the manufacture of swords, knives, surgical instruments, tools, or machine parts. Steel can either be made very tough, or it can be made very hard and ground to a sharp edge. These characteristics have been immensely useful to humans both before and after the Industrial Revolution.

However, as we showed in Chapter 3, the problem of making a compact, highly loaded structure is different from that of making a diffuse, lightly loaded one. Important as tools, weapons, and machinery are, diffuse structures such as houses, aircraft, truck bodies, containers, and furniture are more widespread and their manufacture consumes far more money and effort. Broadly speaking, steel is not a suitable material for these kinds of structures.

In diffuse structures there is usually no special need for the material to be very hard. The problem of supporting tension loads is normally a minor one, but the weight cost of supporting compressive and bending loads can be a major difficulty. Although steel is dense, its tensile strength can be

ALUMINUM

Above, Pinin Farina's sleek 1946 Cisitalia 202 GT has an aluminum body. Below, a nickel-aluminum superalloy of hard crystals embedded in a softer matrix. The regularity of the crystal makes it highly resistant to dislocations, in contrast to the unordered molecular arrangement of the matrix, which is susceptible to dislocations. The subdivided structure of the superalloy incorporates the virtues of both component alloys: it is tough as well as hard, and very resistant to cracking.

Among the metals, aluminum is nowadays probably next in importance to iron and steel. It readily forms on its surface an oxide film that protects it from corrosion. Its electrical and thermal conductivity is high, making it a suitable material both for cooking vessels and for electric cables. Furthermore, pure aluminum is very ductile, so aluminum foil is used very widely for packaging.

Although pure aluminum is very soft and weak, when it is alloyed with other metals such as copper and magnesium it can show excellent combinations of strength and toughness. Structurally, the attraction of aluminum alloys is based on their low density. The specific gravity of aluminum alloys such as *duralumin* is approximately 2.8, which is just over one-third of that of steel at 7.8.

Compared with steel, the Young's modulus and the available combinations of tensile strength and work of fracture are, weight for weight, very much the same. Thus for purely tensile applications there may be little or no advantage in using aluminum, especially when we consider that most steels have definite and predictable fatigue limits whereas aluminum alloys do not.

The real advantage of aluminum arises from its behavior in compression, especially in lightly loaded structures. For panels and columns aluminum and its alloys are about twice as efficient as steel (see the table on page 78). Because thin panels such as the skin plating of aircraft wings and fuselages tend to wrinkle in shear (due to the compression component of the shear stress; Chapter 2), the use of aluminum alloys rather than steel may nearly halve the structure weight of an airplane. When we add the fact that thin aluminum alloy plating is much more resistant to corrosion than steel plating, the case for aluminum in most aerospace applications becomes clear.

Similar arguments in favor of aluminum apply to structures like yacht hulls and car bodies—although here we are apt to run against the problem of cost. Aluminum is made from its ore, bauxite ($Al_2O_3 2H_2O$), by an electrolytic process requiring over four times as much energy, per unit weight, as the manufacture of mild steel. Furthermore, electricity is often more expensive, per joule, as a source of energy than coal. With energy likely to get increasingly costly in the future, the commercial prospects of aluminum as a material for everyday structures seem uncertain.

made roughly equal, weight for weight, to that of most other materials. Again, its stiffness, as measured by its Young's modulus, *E,* is much the same, weight for weight, as that of most competitive materials (see the table on page 78). However, as we discussed in Chapter 3, the efficiency of a column in compression depends, not upon E/ρ (where ρ is the density), but upon $\sqrt{E}/\rho$. The efficiency of panels, which are even more ubiquitous in technological structures than struts or columns, depends upon $\sqrt[3]{E}/\rho$. By these criteria steel is one of the least efficient of materials. It is much less efficient than wood or aluminum, and its does not even compare particularly well with brick or concrete.

It is true that the disadvantages of steel in lightly loaded structures can be offset to some extent by making corrugated iron panels for such functions as roofs, and by shaping columns and joists into tubes and I beams. However, there are limits to the effectiveness of these circumventions: they do not reduce the risk of local buckling in thin-walled sections, nor do they eliminate the liability of mild steel to corrosion.

Another way of maximizing the usefulness of steel is to use it in conjunction with concrete. Steel is a good material in tension, whereas concrete is fairly good in compression. In reinforced concrete the tensile loads are taken mainly by the steel reinforcing rods, whereas the compressive loads are supported partly by the concrete, which in any case stabilizes the steel rods against buckling and tends to preserve them from rust. Although reinforced concrete has been very successful—large buildings, bridges, dams, and even yachts have been built from it—it is essentially a heavy material and normally limited to use in rather ponderous structures.

Within its limitations, steel has one shining virtue: it can be modified fairly widely. Ordinary carbon steels have carbon contents that may vary from about 0.1 to 0.8 percent. Although their mechanical properties will depend to some extent upon the heat treatment they have received, low-carbon steels are generally tough but rather weak in tension, whereas the higher-carbon steels are generally stronger and harder but more brittle. In other words, the same material cannot be both very strong and very tough. A typical low-carbon mild steel might have a tensile strength around 60,000 p.s.i. (400 MN/m^2) with a work of fracture of about 5.5×10^{-5} J/m^2. A high-tensile plain carbon steel might reach 200,000 p.s.i. in a tensile test but its work of fracture is likely to be very low, between 10 and 100 J/m^2. Better combinations of strength and toughness can be achieved with some alloy steels, but, as we said, these steels are more expensive.

The crucial consideration is the *scale* of the component or the structure one is trying to make. The length of a critical Griffith crack is an *absolute,* not a relative, dimension (Chapter 4). Thus, small parts can be made safely from high-tensile steel that might be very dangerous in a large structure. Strong, brittle steels can be used for small tools, springs, and subdivided

Photomicrograph of concrete.

RUST

Above, the Picasso sculpture in Richard J. Daley Plaza, Chicago. Rust may have aesthetically satisfying effects. Right, how rust forms. A drop of water on an iron surface allows the formation of dissolved $Fe^{2+}(aq)$ ions.

For various reasons modern commercial mild steels tend to rust rather quickly, much more quickly than the traditional wrought iron of the nineteenth century. With the ever-increasing cost of labor, the maintenance of a steel structure, especially in a marine environment, can be a serious consideration. Rust can be prevented or delayed by *galvanizing* the steel: coating it with another metal such as zinc. But this practice again costs money and may introduce complications such as unwanted electrolytic action between the metal coating and brass or copper parts.

Stainless steels are alloys of iron that usually contain chromium and nickel. Many of these materials are largely proof against ordinary corrosion and have fairly high tensile strengths, but they are generally too expensive to be used in large structures. Moreover, most stainless steels are not particularly tough and their fatigue properties are not reliable.

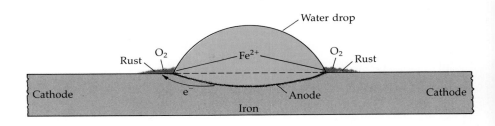

structures such as wire ropes. For large components, especially panels like ship's plating, it is wise to use a much weaker, tougher material. The same is roughly true for car body panels, boilers, and pressure vessels.

A MATERIAL FOR SINNERS?

The widespread use of steel for so many purposes in the modern world is only partly due to technical causes. Steel, especially mild steel, might euphemistically be described as a material that facilitates the dilution of skills.

As Henry Ford and many other industrialists have found, it is particularly suited to the division of labor. Manufacturing processes can be broken down into many separate stages, each requiring a minimum of skill or intelligence, while at the same time the effects of carelessness are limited. Moreover, a fixed rate of working is imposed on the lazy by the pace of the production line.

At a higher mental level, the design process becomes a good deal easier and more foolproof by the use of a ductile, isotropic, and practically uniform material with which there is already a great deal of accumulated experience. The design of many components, such as gear wheels, can be reduced to a routine that can be looked up in handbooks.

One consequence has been that managers and accountants, rather than engineers, have become the dominant personalities in large organizations. Creative thinking is directed into rather narrow channels. Steel is, archetypically, the material of big business—of large factories, railroads, and so on. But "small is beautiful," and small is more likely to be creative. In the author's view the technologies to the twenty-first century will not be based primarily on steel. If this is so, the world may become a more exciting place.

The greatest improvement in the productive powers of labour, and the greater part of the skill, dexterity and judgment with which it is anywhere directed, or applied, seem to have been the effects of the division of labour.

ADAM SMITH
The Wealth of Nations (1776)

I could not possibly do the same thing day in and day out, but to other minds, perhaps I might say to the majority of minds, repetitive operations hold no terrors. To them the ideal job is one where the creative instinct need not be expressed.

HENRY FORD
My Life and Work, Chapter 7 (1922)

Bethlehem, Pennsylvania, 1935. Large steelworks dominated the skylines of the industrial world in the late nineteenth century and most of the twentieth.

6

ANIMAL SOFT TISSUES

The Problems of Life at High Strains

The young painter must become the patient pupil of nature, he must walk in the fields with a humble mind. No arrogant man was ever permitted to see nature in all her beauty—the art of seeing nature is a thing almost as much to be acquired as the art of reading the Egyptian hieroglyphics.

JOHN CONSTABLE, R. A. (1776–1837)
"Lecture on Painting"

Foxglove, Digitalis purpurea. *Though rigid materials are more common in plants, soft tissues are very important, especially in the early stages of growth.*

Metals, especially iron and steel, have played an important and conspicuous part in shaping, and perhaps in dividing, the modern world. But, by their very success, they have also had a rather unfortunate effect in shaping ways of thought among both experts and laymen. On the one hand, the world of "serious" engineering has been preoccupied with rigid structures made from metals, concrete, masonry, and so on. Most traditional engineers have tended to look down on "soft" structures, and they simply have not concerned themselves with things like worms and tadpoles, however ingenious their construction. On the other hand, the world of doctors and biologists has devoted itself to the study of soft tissues of various kinds. Although these people have definitely been interested in worms and tadpoles, their interest did not, until quite recently, extend to the mechanical properties of such creatures. For the most part, they did not know, or want to know, about the engineering principles governing the activities of these living structures. Both sides of the divide have displayed something of the arrogance of which Constable, one of the greatest painters of nature and landscape, complained.

Of course, doctors and biologists have given a certain amount of attention to the strength of bones, partly because bones are apt to break and partly because, as rigid structures, some elementary engineering ideas obviously applied to them. On the whole, though, the mechanical properties of soft tissues have been taken for granted. Their strength and elastic behavior have been considered incidental to their growth mechanisms and metabolic functions, and have not usually been thought to present any very difficult or interesting problems.

However, there is every reason to suppose that Nature has been quite as subtle in designing structures as in developing chemistry and cybernetics. Even the simplest plants and animals are subject to various mechanical loads, and unless they are structurally adequate they cannot exist.

Unlike human engineers, Nature has made little or no use of preexisting inorganic solids, but has apparently preferred to evolve, de novo, organic load-bearing materials especially suited to the purpose. The great majority of these are soft and highly extensible. In animals, rigid materials such as bone, horn, tooth, and shell appeared fairly late in the evolutionary program, and they generally constitute only a small portion of the animal. In plants, rigid materials such as wood are more common, but even in plants soft tissues are very important, especially in the earlier stages of growth.

A living animal made almost wholly from rigid materials would have to resemble an engineer's machine. It would presumably work by means of mechanisms like axles, gears, crankshafts, and pistons. It might even prefer wheels to legs. The mind boggles in contemplating how such creatures could evolve, or even reproduce themselves. But Nature is much cleverer than that: the development of living things is based, to a very large extent, upon the exploitation of soft, extendible tissues that are capable both of carrying the necessary loads and of growth and evolution.

Because they have high Young's moduli, the rigid materials to which the engineer is addicted can resist considerable forces while being deflected only by a small amount. Such materials generally operate under their working loads at elastic strains—that is, strains from which the material can recover—in the region of 0.1 percent; rarely, they might strain to 1.0 percent; but they almost never operate at higher strains. However, many of the ordinary tissues of the human body commonly work at fully elastic strains between 50 and 100 percent, and these strains are, of course, fully elastic and recoverable. Sometimes living tissues work at much higher strains than this: the cuticle of the pregnant locust, who is in the habit of depositing her eggs at the bottom of a deep, narrow hole, can extend 1200 percent, and still recover elastically!

It is true that the working stresses developed in most animal soft tissues are lower than those to which many engineering structures are subjected. But this does not mean that the design problem is any easier in an animal than it is in an aircraft. Because the strains are so large, the strain energy stored per unit volume of stressed biological material is generally at least equal to that in a critical engineering structure; in the case of animal tendon, the stored energy may be many times higher. Thus, from the point of view of fracture mechanics, the problem of the safety of living soft structures is anything but a trivial one, and it is generally true that Nature is more successful in solving her fracture mechanics problems than are modern engineers.

A structural system based on strains that are, roughly speaking, 1000 times greater than those to which conventional engineers are accustomed must differ in many important aspects from classical engineering. Not many scientists were prepared to risk plunging into these rather deep and murky waters. However, in 1940, the British scientific establishment extracted a Cambridge zoologist named Mark Pryor (1915–1970) from the British army and sent him to do the rest of his war service at the Royal Aircraft Establishment at Farnborough—which, it will be remembered, had already given a home to A. A. Griffith, the father of fracture mechanics.

Throughout the rest of the war Mark Pryor and I worked on various aspects of nonmetallic materials in aircraft. Mark made some very important contributions to the construction and the safety of wooden aircraft, notably to the immensely successful Mosquito bomber. He was also much involved with the wooden troop-carrying gliders that made the invasion of Normandy possible. However, amid these serious wartime activities, Pryor found the time to discuss with me, at great length, the application of engineering ideas to biology. This collaboration between an entomologist and a naval architect—if such I can call it, because most of the ideas were Pryor's—must have been unique. Perhaps in consequence of all this we were regarded with some suspicion by the pundits in the Structures Department at Farnborough. In 1945 Pryor returned to Cambridge, where he became one of the original pioneers of the field that is now called biomechanics. I stayed on at Farnborough and continued to battle with the pundits in the Structures Department. Unfortunately, he died in 1970 a few years before his rather eccentric subject became fashionable.

Unknown to us at the time, another pioneer of biomechanics was working on the other side of the world, in Japan. A medical academic, Dr. Hiroshi Yamada, began to study the biomechanics of people and animals at the Kyoto University of Medicine toward the end of 1938. His research was interrupted by his war service as a surgeon in the Japanese army, but he resumed it in 1948. Professor Yamada measured the mechanical properties of an enormous number of living tissues, and his book, published in English in America in 1970, is a valuable standard reference on the subject.

Interdisciplinary pioneers like Pryor and Yamada were fairly exceptional, and neither was much noticed or greatly respected at the time. Only in the last 10 or 20 years has the mechanical behavior of soft tissues been studied widely and quantitatively. Predictably, it has turned out to be more difficult, more complicated, and much more subtle than the philistines had supposed. There is much that we do not yet understand about molecular mechanisms, which is why this is a fairly short chapter. However, due to the now fashionable collaboration between biologists and engineers, we have begun to bridge the divide between "serious" engineering and biology. These days, we have a much clearer view of what some of the structural problems of life really are, even if we are not yet clever enough to comprehend many of the ways in which Nature has chosen to solve them.

ELASTICITY AND SAFETY AT HIGH STRAINS

The closer we look, the more evident it is that the problem of designing soft, extensible structures is different in kind from that of designing traditional engineering structures.

As we said in Chapter 4, the elasticity of ordinary rigid crystalline and glassy solids is based on the direct extension of the interatomic bonds in the material. Although in a conventional safe structure the working strains are nearly always below 1.0 percent, it is possible, by exercising great care, to strain such materials elastically almost up to the full theoretical 20 percent or so before they fracture. Beyond this limit a normal, Hookean solid cannot extend and then recover itself undamaged, because the interatomic bonds can be stretched no farther.

It is possible to conceive of a soft animal or plant cell whose flexibility is limited to strains of less than 20 percent. However, we also pointed out in Chapter 4 that the strain energy stored in a Hookean material when it approaches its maximum theoretical strain of around 20 percent is about equal to the total chemical bond energy, and is therefore comparable to the energy of an equivalent weight of explosive. If the material is broken mechanically at anything approaching the maximum elastic strain, this energy will be released very suddenly and there will be an explosion. To break a flexible plant or animal made in this way, for instance by eating it, would therefore be equivalent to putting a match to a stick of dynamite!

Although ductile metals and "plastic" materials like soft clay can be made to extend 100 percent or so without any dramatic or dangerous consequences, such extension is, of course, an inelastic process. If we want the material to recover its original shape, we shall have to push it back again. If we do this more than a few times, the material will break. Plastic extension of this sort is little used in biology.

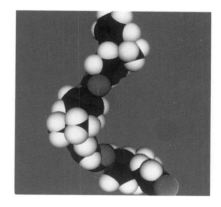

Rubber is a natural polymer. Polymer molecules form a long chain whose structure is similar to that of the artificial polymer LEXAN. Here, blue spheres indicate the carbon backbone, red spheres oxygen atoms, and white spheres hydrogen atoms.

In technology, the longest known and most familiar highly extensible elastic material is rubber. Both natural and synthetic rubbers can often elastically about 800 percent, which is higher than the extensions of most animals. Moreover, rubber can be stretched and relaxed many times without suffering much damage. It could be supposed that rubber might therefore present a model for soft biological tissues, but, broadly speaking, this is not true. Although natural rubber is made from the latex of the tree *Hevea brasiliensis*, botanists say that there is no evidence that rubber, or any similar material, has any structural role in the rubber tree or in any other plant. Rubber is, so to speak, a biological accident or freak.

Why are rubberlike materials not suitable for the majority of natural soft structures? The answer is that they have the wrong sort of elasticity. Both natural and synthetic rubbers depend for their extendibility on their long, flexible molecular chains. In the contracted or resting state of the material, these chains are coiled upon themselves, or sometimes just packed together, like a bundle of string. The strong primary bonds operate only along the backbone of the chain, but by virtue of secondary bonds such as van der Waal's forces, the chains cling to each other laterally in a weak way. When the rubber begins to be strained in tension this lateral adhesion offers a certain amount of resistance to the disentangling of the chains, so the stress-strain curve rises fairly steeply to begin with.

Once most of the chains are separated, however, the chief resistance to further extension lies in the would-be thermal or Brownian motion of the chains; the random thermal forces are tending to make the flexible chains

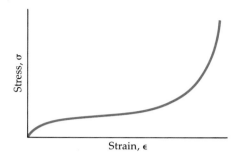

The stress-strain curve for a rubbery polymer is sigmoid or S-shaped.

kinked or convoluted. It therefore requires a tensile stress to extend and straighten them, but this stress does not increase very rapidly with extension, so the stress-strain curve now rises less steeply. Eventually one reaches a point where most of the chains have been straightened, so that one is pulling on the primary chemical bonds in the backbone of the chain. These bonds are very stiff, so the stress-strain curve now bends upward again rather steeply. The result is a sigmoid or S-shaped curve. The theoretical curve agrees very closely with large numbers of experimental results on natural and artificial rubbers.

One has only to stick a pin into a blown-up child's balloon to realize that ordinary rubber is a thoroughly dangerous material to use in a structure, especially for things like the stressed membranes that are so common and so important in plants and animals. Rubber behaves in this brittle way, partly because the sigmoid shape of its stress-strain curve encourages concentrations of stress at any hole or irregularity, and partly because unreinforced rubber has a low work of fracture. Although it is possible to render rubber tough by incorporating fibers in it, as the tire manufacturers do, most of its extendibility is lost this way.

Another serious objection to the biological use of materials with a sigmoid stress-strain curve in biology is their tendency to form aneurysms. If a material like this were used to make a tube such as a vein, an artery, or a piece of intestine, it would be liable to bulge out under internal fluid pressure into the sort of local swelling that doctors call an aneurysm. The mathematical analysis is a little tedious, but we can demonstrate the phenomenon by blowing up a cylindrical child's balloon.

Aneurysms may be formed in a balloon (left) or, more seriously, in blood vessels. Right, X-ray photograph of a large lesion in an elongated aorta. The thin white line is a catheter through which a medium opaque to X-rays was introduced into the aorta. Aneurysms may fail catastrophically (i.e. burst); for an example, see the illustration on page 74.

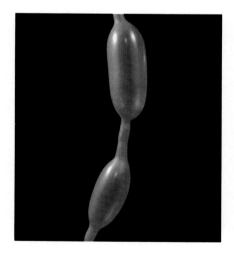

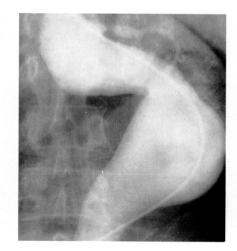

CHAPTER SIX

Because both animals and plants can be regarded as systems of tubes containing fluids, such as blood and sap, under pressure, the tendency to form aneurysms precludes the use of this kind of elasticity in most living things. Indeed, a biological sigmoid stress-strain curve seems to be found only in human and animal hair, where, presumably, it cannot do much harm.

THE J CURVE

The model or archetype for a great many biological soft tissues is a liquid— or rather, the surface of a liquid. Liquids have surface tension that is different in important ways from tensile stress on a solid. For one thing, surface tension is not affected by the depth or "thickness" of the liquid; but more significantly for our present purpose, the tensile force in a liquid surface is not increased by its extension. In other words, the stress-strain "curve" is flat.

Resistance to the extension of a liquid surface is caused, not by an increase in the surface tension force, but by the increase in energy that must occur as the area of the liquid surface gets larger. This sort of "elasticity" has several consequences. For one thing, since the stress is constant, there can be no stress concentrations. When a bubble bursts, the fracture mechanics responsible are quite different from those governing the behavior of an ordinary solid. The critical factor is surface area. A very small hole in a bubble wall will cause an increase in surface area and hence in surface energy; such a condition is self-healing because the system will favor a return to the lower surface energy of the intact bubble. If the hole is enlarged beyond a certain critical diameter, however, the total area of the system will be diminished, so the hole will expand and the bubble will burst. With a deeper liquid such conditions do not apply, and it is impossible to fracture, say, a pool of liquid.

In many ways the surface of a liquid is therefore well suited to form the boundary of a living cell. It is *possible*—though not proven—that the earliest forms of life were contained in droplets of a liquid, presumably aqueous. The surfaces of such "cells" would have been self-healing, but if they grew and swelled sufficiently the droplets could have broken in two, rounded themselves out again, and so multiplied. Liquid emulsion systems make an appealing hypothetical cradle for the most primitive forms of life—but most of this is speculation.

What is certain is that the elastic behavior of the soft tissues of both primitive and advanced animals at low and moderate strains greatly resembles the surface of a liquid. After such tissues undergo a certain amount of extension (with the amount varying between different tissues), the stress-

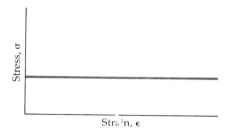

The stress-strain curve for surface tension in a liquid.

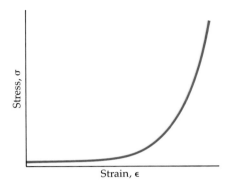

The J-shaped stress-strain curve (or simply J curve) that is typical of nearly all soft animal tissue. It results in mechanical toughness and discourages aneurysms.

strain curve tends upward rather steeply. The result is the J-shaped stress-strain curve (see margin at left) curve that is characteristic of many soft tissues which, as we shall see, offers them some important advantages. As is also true of the surface tension of a liquid, the curve does not really pass through the origin of the coordinates, because the stress on living tissues is never zero. In vertebrate animals especially, most of the soft parts are pre-stressed in one way or another; that is, in their neutral or resting position they are still under tension, like the rigging of a ship.

Before discussing the molecular mechanisms that might be employed to produce this sort of elasticity, it is worth considering the benefits that the J curve has to offer. When biomechanical engineers became aware of the extreme toughness of many biological membranes, such as human arteries, they at first supposed that the work of fracture of the materials involved must be very high. However, when the author's colleague Dr. Peter Purslow measured the works of fracture of a considerable number of animal tissues, they were not exceptionally high, being usually in the range of 10^3 to 10^4 joules per square meter. The toughness of rat skin, worm cuticle, or human arteries differs from metal sheet, or indeed from rubber sheet, not so much in the work of fracture as in the shape of the stress-strain curve.

Because the lower part of the J curve is virtually horizontal, the material can have little or no shear modulus in this region. There is thus no effective mechanism whereby the released strain energy can be transmitted to the fracture zone: an elementary but elegant safety mechanism. Furthermore, the mathematical characteristics of the J curve are such that, unlike those of the sigmoid or S curve, tubular pressure vessels such as arteries will not normally develop elastic instabilities such as aneurysms. The J-curve approach to fracture mechanics is now being examined by engineers.

The toughness of animal skins is reflected in the history of technology by the widespread use of leather and rawhide. The human race would have progressed a good deal more slowly without leather sandals and shoes, not to mention horses' harness and a great many other tough, flexible artifacts. Until the recent introduction of synthetic soft materials, there was no good substitute for leather.

To confirm the toughening effect of the J curve we can consider the properties of the shell membrane in the eggs of birds—a material that does *not* possess this sort of elasticity. The shell membrane is the thin, continuous membrane that exists just inside the hard shell of an egg. Because the shell itself is permeable, the function of the membrane is to protect the embryo by keeping moisture in and infections out. When the chick hatches, however, it has to break its way out of the egg, and, of course, the

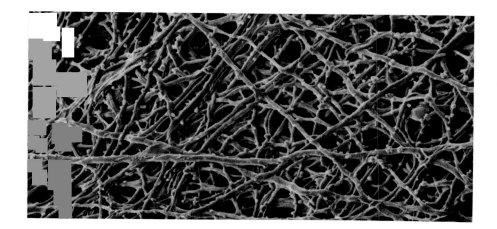

Left, the membrane adhering to the inside of an eggshell is composed of ovokeratin, a highly insoluble protein (magnified 1,200 times). Below left, eggshell lying atop the membrane (magnified 300 times). Below right, a hatching bird—here a Muscovy duckling— must first break through the shell membrane.

baby is not very strong. A tough shell membrane would present a serious obstacle to the young bird.

The shell membrane of eggs is almost unique among animal soft tissues in obeying Hooke's law up to an elastic strain of about 22 percent. The molecular mechanics responsible for this behavior is apparently not known. It is certainly not due to the direct stretching of the interatomic bonds to 22 percent, because the Young's modulus is very low indeed: about 7 MN/m^2 or 1,000 p.s.i. One infers the existence of a most ingenious molecular fudge. What is certain is that these shell membranes tear very easily, which, of course, is just what is wanted.

Outside biology, the world of textiles provides some good examples of the effect of J curves versus Hookean elasticity upon the tearing of "membranes." A knitted fabric has a J-shaped stress-strain curve, as one can easily feel by pulling on a sock or a knitted jersey. Such fabrics are usually hard to tear. Woven cloths more or less obey Hooke's law in the direction of the threads—that is, in the warp and weft directions (see Chapter 2). At other angles—that is, on the bias—the cloths show J curve-type behavior. Most woven fabrics can be torn, accurately and in a straight line, at 0° and 90°—again, in the warp and weft directions. In fact, this is the way in which dressmakers commonly do tear cloth, although these are the "strongest" directions of the cloth. It is very difficult to tear a woven cloth at 45°. Likewise, a casual rent—say in the seat of one's pants—is nearly always L-shaped, with the arms of the L in the warp and weft directions.

Curiously, the human amniotic membrane that contains the fetus and that has to split in childbirth does show a J-shaped curve, though rather a steep one. However, Yamada has found that this membrane is considerably weakened during the last two months of pregnancy; no doubt physiological changes—probably weakening of the chemical bonds in the membrane—take place that are not directly connected with the elasticity of the material. Ordinary human skin can become quite weak and brittle during periods of ill health.

THE MECHANISMS THAT PRODUCE THE J CURVE

Most of the mechanisms of deformation that occur in nonliving materials have been more or less fully understood for a number of years. Classical Hookean elasticity is the consequence of the direct stretching of interatomic bonds. The plastic behavior of ductile crystals, such as metals, is due to the operation of dislocation mechanisms. The sigmoid elasticity of rubberlike solids, which we have described earlier in this chapter, has been mathematically analyzed and accounted for. But how do biological soft tissues work? What molecular and other mechanisms are responsible for the J curve? Until quite recently no very convincing answer had been given. Even now, because the mechanisms of soft tissues can vary greatly in detail and can be complicated, the subject is not by any means free from controversy. But we do now at least understand the broad principles.

There is not much difficulty in explaining the steeper part of the J curve—that is, the region of high stresses and strains. When any system of flexible convoluted fibers or molecules is stretched until its threads are nearly straight, further increases in tensile stress can produce relatively little additional extension.

The difficulty is to understand what is happening near the beginning of the curve—that is, in the nearly horizontal zone. In this region stress is nearly independent of strain: very small stress loads produce relatively large extensions. This regime may continue up to strains of 60 percent and sometimes considerably more. The possession of this sort of elasticity confers very great benefits on most living things. As we have said, such behavior is closely analogous to the surface tension in a fluid, but surface tension is confined to two dimensions. In biological solids the elasticity operates in three dimensions—right through the thickness of the material. I suggested in 1974, and Professor S. A. Wainwright of Duke University has emphasized, that the development of materials exhibiting this sort of elasticity was an evolutionary milestone whose very great importance has not been realized by biologists until quite recently. As far as we can tell, it seems to be an essential requirement for the existence of life as we know it.

Two parallel, coexisting, mechanisms appear to be involved. First of all, we can extend to the molecular level the knitted fabric analogy that we used to describe J-curve behavior: there is no doubt that many complex molecular networks in biology are constructed like knitted fabric. The second mechanism, which is often more important, is quite different and perhaps more subtle.

Recently, Professor John Goseline, of the University of British Columbia, and his colleagues have demonstrated a *hydrophobic* mechanism of elasticity in many soft tissues, especially those containing elastin. (*Hydrophobic* is from the Greek for *water-fearing*.) This system works by the forcible exposure of nonpolar molecular groups to water, which is a highly polar molecule. In a simplified model, molecular chains or networks containing groups resistant to water were extended in such a way that these hydrophobic units were made to come into intimate contact with the polar environment of an aqueous gel. The energy changes could be calculated and the stress-strain characteristics of the system (which might be described as the opposite of a wetting agent) did resemble a three-dimensional version of surface tension: stress was constant and independent of strain.

Moreover, like surface tension, the mechanism is fully reversible; in fact, indefinitely so. The walls of blood vessels, such as arteries, are subjected to around 40 million repetitions of stress (heartbeats) in the course of a year; the satisfactory circulation of blood depends upon the repeatability of the strain mechanisms in the circulatory system over, one hopes, a period of many years. Fatigue failures of the type engineers encounter with metals are fortunately rare in soft tissues.

COLLAGEN AND TENDON

Although elastin is an important constituent in most soft animal tissues, it is seldom used on its own. It is generally employed in combination with stiffer substances that serve to reinforce it and modify its elastic properties. Of these substances, the most notable is collagen.

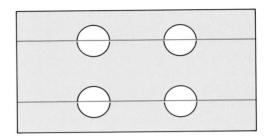
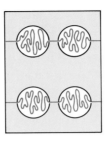

Hypothetical morphology of elastin. Left, in the resting or unextended state, the chain molecules are folded within droplets. Right, in the extended state, the molecules pull out of the droplets.

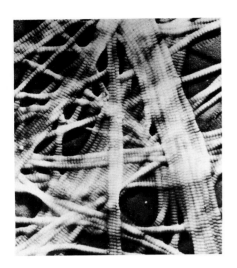

Fibrils of collagen, a polymer. Dark bands spaced approximately 700 angstroms apart appear where the intricate structures of the collagen molecules overlap.

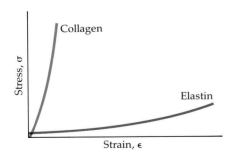

Stress-strain curves for elastin and collagen.

Collagen is essentially a fibrous material constructed from long polypeptide chains arranged to form a series of rather steep helices whose geometry is not very different from that of ordinary string. In fact, collagen behaves in tension much as one would expect a stringlike substance to behave. Initially the helix contracts upon itself; that is, it reduces its diameter somewhat. During this early stage it extends relatively easily, though not nearly as easily as elastin. However, as the twisted fiber contracts in diameter it hardens and stiffens so that additional extension at the higher stresses is due to the primary bonds along the backbones of the polypeptide chains. The result, elastically, is a J curve of sorts, but a much steeper one than that shown by elastin.

Clearly, by combining collagen, elastin, and other soft components such as muscle tissue, a wide range of elasticities can be produced. Furthermore, although the work of fracture of most animal tissues is not exceptionally high, it is by no means negligible. Most of the work of fracture in materials such as tough meat is due to the presence of collagen fibers, which function somewhat like the fibers that are used to reinforce the walls of automobile tires. When meat is cooked, the collagen disintegrates at a lower temperature than the elastin and the other constituents, which is why meat, hopefully, becomes tender when it is cooked at temperatures around 100°C. Not surprisingly, the biomechanical phenomena involved in tenderizing meat are of considerable commercial interest. In fact, far more of the funding of research on the works of fracture of animal tissues comes from the meat industry than from medical, academic, or government sources.

To the engineer, perhaps the most interesting aspect of collagen is its use in the construction of animal and human tendons. Tendons perform two functions. First, they serve to transmit the tensile forces generated by the muscles to the limb bones and other parts of the body to make movement possible. It is fundamental to the engineering design of vertebrate animals that muscles be able to operate, as it were, from a distance. The human hand, for instance, would be a clumsy device if all of its muscles had to be attached directly to the bones of the fingers. In fact, most of the muscles that operate the hand are situated quite a long way up our arms; the muscles are connected to their relevant bones by means of long, thin tendon cords. It is easy to feel their operation with the fingers of one hand by touching the various parts of the other hand, wrist, and lower arm while wriggling them.

Tendon has another function besides the direct transmission of tensile forces from the muscles to the bones. It also serves as a store for strain energy—that is, as a spring. One has only to watch deer and horses leaping, cats pouncing on mice, squirrels and monkeys bounding along tree limbs, and so on, to realize that many animals are equipped with very efficient storage mechanisms for strain energy.

Tendons consist predominantly of parallel collagen fibers arranged in a sort of skein. Compared with most other soft tissues, tendon is a compara-

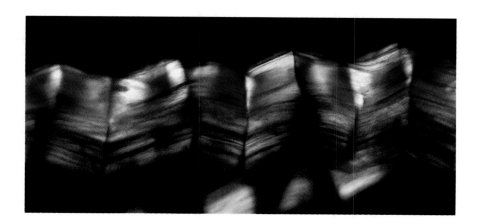

A tendon is made up of bundled fibrils of collagen. Polarized light reveals the crimping of the fibrils, which makes them elastic. (For an explanation of polarized light, see page 14.)

tively stiff material, as indeed it has to be to transmit the muscular forces precisely. Its maximum extension is rather low by biological standards, usually between 8 and 10 percent. On the other hand, its capacity for storing strain energy is exceptionally high—much higher than that of any of the traditional engineering materials.

Approximate strain energy storage capacity of various solids

Material	Working strain (%)	Working stress lb/in^2	Working stress MN/m^2	Strain energy stored (J/m$^2 \times 10^6$)	Specific strain gravity	Energy stored (J/kg)
Ancient iron	0.03	10,000	70	0.01	7.8	1.3
Modern spring steel	0.3	100,000	700	1.0	7.8	130
Bronze	0.3	60,000	400	0.6	8.7	70
Yew wood (in bending)	0.9	18,000	120	0.5	0.6	900
Tendon	8.0	10,000	70	2.8	1.1	2,500
Human hair*	30.0	23,000	160	14.0	1.1	13,000
Horn	4.0	13,000	90	1.8	1.2	1,500
Rubber	300	1,000	7	10.0	1.2	8,000

*Hair has only a 60 percent recovery after stretching; it also creeps badly under stress. It is therefore not as well suited for the storage of strain energy as the figures for stress and strain might make it appear.

One might suppose that this capacity for storing impressive amounts of strain energy would lead animals into problems with fracture mechanics, but Nature dodges this threat by good design. In the case of tendon, the strategy is subdivision. First of all, the tendon itself consists of a large number of parallel strands of collagen fiber that are only weakly attached to one another laterally. Thus the Cook-Gordon mechanism (Chapter 4) nearly always prevents a crack from spreading across the system. Secondly, the ends of the tendon are attached to the bone and to the muscle by many small separate joints. A few of these joints can fail without catastrophic results. Very similar arrangements used to prevail in the design of the shrouds of traditional sailing ships, but they are by no means always employed by modern engineers.

Actually, there is in animal tendon a third safety mechanism that non-living structures have not been able to utilize: animals have a sense of pain. When we "pull" a tendon we feel intense pain, although we have nearly always broken only one or two out of the many joints by which the tendon is attached to the bone. In the engineering sense the system no doubt retains a considerable factor of safety, but we are usually careful not to put much load on it until it has fully healed.

Engineers have not yet given much attention to the efficacy of providing the equivalent of a sense of pain in a technological structure. However, it might not be beyond the capacity of modern electrical and electronic engineering to devise a surface coating that when applied to critical structures would be sensitive to the presence of incipient cracks and operate some kind of early warning mechanism.

BOWS AND CATAPULTS

The figures given in the table on page 150 for the storage of strain energy in various materials shed some light on the history of weaponry. We do not know when arrows shot from bows were first used for hunting or warfare, but it was certainly fairly early in the history of technology. Moreover, the bow seems to have been invented independently in many different countries and continents. The material of choice for the strain energy storage element—that is, the main body of the bow—was wood.

The English longbow, which may have changed the course of history at Crécy and Agincourt (Chapter 5), was made from yew, which has various advantages for the purpose. It is often supposed that the English bows were made from the yew trees which grow in so many English churchyards. In fact, most of the English bowstaves were imported, together with shiploads of wine, from Spain. Although yew grows prolifically in Spain, Italy, and other Mediterranean regions, bows made from yew or any other wood were seldom used in southern countries. This was because yew, and other wood to a lesser extent, begin to creep badly under stress at temperatures above 35°C. It may have been partly for this reason that English armies did not cross the Alps or the Pyrenees during the Middle Ages, but confined their depredations to France and the Low Countries.

Bows made from bamboo were used in various parts of the Orient. They are still in use in rather remote places like the Himalayas. I do not know how effective they are.

Because of the limitations of bows made from plant materials, *composite* bows constructed in various ways from animal tissues were developed in various countries from early times. In particular, the ancient Persians used composite bows, probably dating from well before 500 B.C. These were the bows used against the Greeks at Marathon in 490 B.C. and, no doubt, against other enemies. The Persian composite bow was a sophisticated construction that, compared with the wooden longbow, must have been expensive to make.

In the Persian bows the forward, or tension, face was of animal tendon, whereas the rear, or compression, face was of horn—another very efficient material. Apparently such bows did not deteriorate in hot weather, and my colleague, Dr. Henry Blyth, has found evidence that unlike wood or metal bows, they could be left strung for long periods without losing any elasticity. Similar composite bows remained in use by the Turks and others for well over 1,000 years until they were superseded by firearms in comparatively modern times.

Although the traditional bow is a light and convenient weapon, the energy that can be stored in it is limited by the archer's strength (a typical maximum force is 50 lb or 20 kg) and by the length of the archer's arms, which restrict the "draw" to about 2 feet or 60 centimeters. Even with intelligent design, it is difficult for the archer to store more than about 170 joules of energy in his bow.

One way of dealing with this limitation is to make a crossbow. A bow that is much stiffer than a conventional bow is fixed across a longitudinal

Tyrolese crossbow made around 1556. The bowstring is hooked by the cranequin, which is pulled back by turning the toothed wheel, drawing the bowstring taut. When the trigger (below the stock) is pulled, a release nut pushes up on the cranequin, releasing the bowstring.

rod or staff that looks something like the barrel and stock of a rifle. The string of the bow can then be drawn by means of a system of levers or gears so that the draw force can be greatly increased. When the bow has been drawn, the string can be held and then released by a catch-and-trigger mechanism somewhat resembling that of a modern gun. Crossbows were fairly widely used in medieval warfare. In fact, they were so effective against armored knights and soldiers that one of the popes forbade their use against Christian opponents; they were to be used only against "heathens."

Actually, the introduction of the crossbow in the late middle ages was more of a technological revival then a new invention. The crossbow, or something very like it, was invented and used to a limited extent by the ancient Greeks. Both the Greeks and the Romans took the development of strain energy weapons even further than this in constructing and using very large and efficient catapults in siege warfare.

The earlier catapult-type artillery consisted of much enlarged crossbows supported by a stand or mounting. In scaling up the crossbow beyond a certain point, the designers must have run into various difficulties. So, instead of storing the necessary strain energy in a beamlike device such as a scaled-up bow, the Greeks developed twisted skeins of elastic fibers, usually animal tendon.

The most effective design of this type of weapon was known to the Greeks as a *palintonon,* and to the Romans as a *ballista.* A few years ago, in a joint project carried out by some classical scholars and engineering students at Reading University, England, a Greek palintonon was reconstructed according to specifications derived from literary sources and modern engineering skills. When it was finished and tested, its performance fell considerably short of the known range and destructive capacity of its historical prototypes, but even so it was an impressive weapon.

Rather surprisingly, since the cow is not a notably active or nimble animal, ox tendon has been found by Yamada and other investigators to have a very high capacity for storing strain energy. In fact, ox tendon was the fiber preferred by the Greeks and Romans. In classical Greece, however, beef was a luxury and ox tendon was scarce. With the decline of the Greek democracies and the rise of the "tyrannies" or dictatorships, the standard of living seems to have risen and tendon became an important munition. In the course of the political wheeling and dealing of the time, dictators and other rulers would send shiploads of tendon to their allies or to those whom they hoped to influence. Modern analogies may occur to the reader.

Although tendon was undoubtedly the best and most effective material for catapult springs, it was not always available. We mentioned that these weapons were chiefly used in siege warfare, which involved both

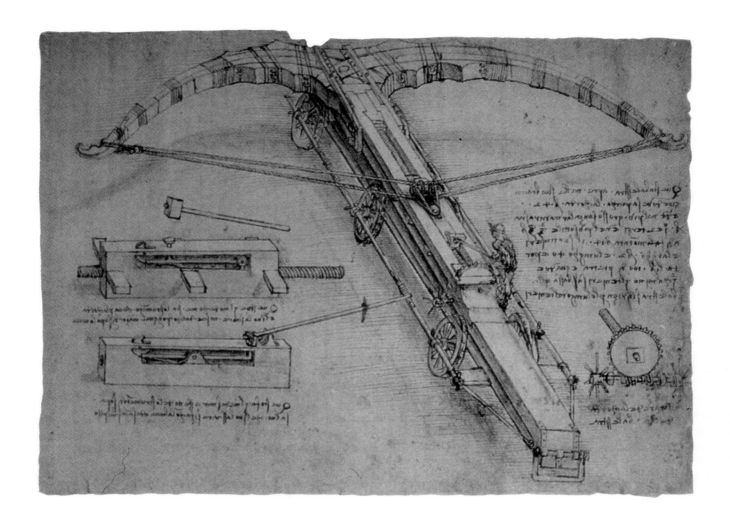

The ballista, an oversized crossbow used from ancient times to hurl large objects, such as stone, was still employed in warfare in the sixteenth century, when Leonardo da Vinci analyzed its tension and compression forces.

attackers and defenders. In other words, catapults were used to batter down the walls of the besieged town, but they were also mounted on those walls in hopes that they could destroy the enemy's artillery.

In a besieged city animal tendon was likely to be in short supply, and the defenders had to fall back on human hair as a substitute. In classical times, men were generally short-haired, so the women would have been called upon to sacrifice their hair. There are many stories about this, some of which are probably legendary. The idea even spread into mythology, for the Romans cultivated the myth of a bald but patriotic Venus. It does seem to be true, however, that during the final catastrophic siege of Carthage in 148 B.C. the women of the city did sacrifice their hair—unfortunately to no effect.

Hair was truly a material of last resort. Although the stresses and strains given for hair in the table on page 150 are impressive, hair creeps badly under load and cannot store much strain energy for any length of time. This condition enables women to use curlers and similar devices for putting a wave into their hair. Presumably the mechanical properties of hair are to a large extent fortuitous, because hair was never designed by Nature as way of storing mechanical energy.

MUSCLES AND ACTIVE ELASTICITY

Thus far we have been dealing with what might be described as *passive elasticity;* that is, with systems and materials whose stresses and strains are induced entirely by the action of external forces. But animals have muscles, which possess two entirely different kinds of elasticity. When muscular tissue in a passive or relaxed state is subjected to a tensile load, it responds, much like other soft animal tissues, by producing a rather flat J-shaped stress-strain curve. However, when muscle is stimulated by an appropriate nerve signal it contracts in length, often by a considerable amount, and so exerts tensile forces in an *active* way.

Such contractions are sometimes conscious or voluntary, and sometimes unconscious, involuntary, or automatic. Voluntary muscular contractions working through tendon mechanisms enable animals to walk, run, swim, and fly. They enable us to use our hands, our jaws, and so on. Contractions that are usually involuntary and automatic serve to operate our hearts, lungs, and bowels. Of these processes most people are well aware.

We tend to be less aware of the active structural functions of muscles that often operate unconsciously to enable people and animals to stand up, for instance. Judged as a passive Euler column in an engineering sense, a human is an inadequate structure that would buckle and collapse under its own weight as soon as it tried to stand upright were it not that any tendency to buckling is corrected at once by a suitable muscular contraction. If this system of active elasticity is interrupted by alcohol, fainting, or death we fall to the ground in short order. There is much to be said for the active muscular approach to countering inherent structural instabilities—and not only in giraffes.

Different kinds of muscles differ in their morphology, but most tend to be built up from a large number of fairly short parallel filaments that overlap one another by a variable proportion of their length. When the muscle contracts or extends, the neighboring filaments slide over one another by an appropriate amount. The mechanism by which they do this seems to be closely analogous to that of the edge dislocation operating in inorganic crystals (Chapter 5). When the muscle extends, the operation is a passive one, like that in the crystal. When the muscle contracts, however, energy is fed into the system so that the "dislocations" occurring in the rows of links between the filaments are enabled to "climb" against the forces that are tending to extend them in shear; and so the muscle shortens.

Muscular action is an energy conversion process; under suitable conditions it can be a very efficient one. The overall efficiency of a fit person riding a bicycle—that is, the proportion of chemical energy contained in her food to her output of mechanical energy in the bicycle—has been measured as around 30 percent. Some workers have claimed even higher figures than this for the metabolic efficiency of animals. In any case, 30 percent is higher than the overall efficiency of, say, a steam or diesel engine.

Optimum efficiency is only achieved, as a rule, when the muscles are contracting fairly slowly. This is why running is so much more tiring than walking. Animals that have to perform sudden rapid movements such as leaping generally store most of the necessary energy in advance in tendon springs, a little like Greek catapults.

For many applications, muscles need to be *geared*. The simplest form of gearing is to assist a muscular function by holding a stick, club, hammer, or tennis racquet. Kinetic energy is stored in the moving mass of the weapon,

Muscular action is an energy conversion process.

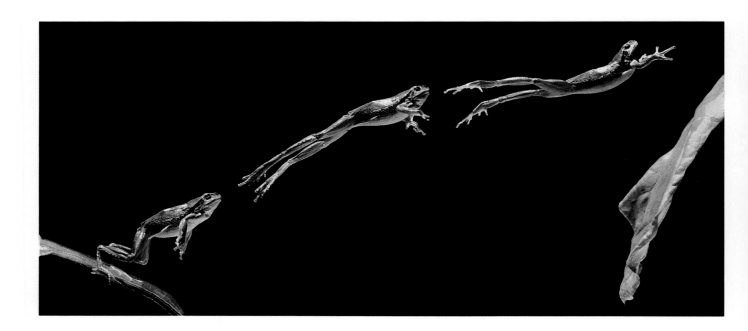

and its length and leverlike form enable the working end to move much faster than the hand that operates it. Moreover, the movement is energetically more efficient than direct action—such as driving in a nail with stone instead of a hammer—and so is less tiring.

An important application of the principle is in rowing. Most primitive boats and canoes were propelled by paddles where the gearing ratio to the muscles was very low. Oars seem to have been invented by the ancient Egyptians. The use of oars enables a much higher and a much more precise ratio between the rate of contraction of the muscles and the speed of the boat to be chosen and used. In fact, this ratio proves to be a very critical one. Modern competitive oarsmen choose the exact lengths of their oars with care and circumspection.

The invention and the exploitation of the oar made the invention of the warship and, with her, the modern concept of sea power practicable. Previous to the development of the galley, ships and boats were used, offensively, only for piracy or else as troop transports. The Greeks and the Phoenicians developed the fast multi-oared war galley. The earlier form was the *penteconter,* which was propelled by fifty oarsmen rowing in a single bank. Sometime around 600 B.C. the penteconter was superseded by the *trieres* or *trireme,* a larger and faster vessel using three superposed banks of oars and 170 oarsmen. Incidentally, in classical times these oarsmen were not slaves but free men who were rather highly paid.

These fast and highly maneuverable galleys were armed with formidable bronze rams enabling them to bar the seas to fleets of troop transports, much as Nelson's fleet barred the seas to Napoleon's armies. The most decisive galley battle in ancient times was at Salamis in 480 B.C. A Greek fleet of triremes defeated a Persian fleet under the tyrant Xerxes; the consequent Greek command of the seas in the eastern Mediterranean made classical Greece secure from Persian aggression throughout the magnificent and creative fifth century B.C.

The researches of J. S. Morrison and other archaeologists have shown that the rowing arrangements in these Greek warships were highly sophisticated. The muscular gearing ratio, as measured by the length of the oars and the position of the rowlocks, was very similar to that used in modern university eights.

It is a little surprising that although the advantages of optimizing the rate of muscular contraction and extending the range of muscular movement were realized and exploited at sea very early in history, it took thousands of years for the same idea to penetrate to the designers of land transport. The bicycle was developed only during the latter part of the nineteenth century. The underlying principle of the bicycle is the same as that of the oar. Just as the length of the oars is important in rowing, so is the gear ratio critical in a bicycle or tricycle.

Reconstruction of a Greek trireme propelled by muscle power. (For the structure of the trireme, see page 13.)

7

SOME STIFF BIOLOGICAL MATERIALS

Wood, Bone, and Antlers

A fool sees not the same tree that a wise man sees.

WILLIAM BLAKE (1757–1827)
"The Marriage of Heaven and Hell"

The engineering of Nature is predominantly an engineering of soft tissues, however, in an intensely competitive world there are advantages to be gained by exploiting more rigid materials, for aggressive purposes as well as for defensive armor. The uses of teeth and claws are obvious. But teeth and claws are not likely to be effective unless they are backed up by a more extensive rigid structure: jawbones, limb bones, and so on.

Why have soft materials not been adapted for purposes of offense and defense? It is true that soft materials can be blown up by fluid pressure in such a way that they can support compressive and bending loads quite efficiently, but these loads generally have to be diffuse ones. If there is a need to apply or resist localized forces, a system based on soft materials is apt to be too clumsy, flexible, and imprecise for many applications. This is especially so when we have to deal with a moving cantilever such as a limb. The arms of an octopus do very well for their marine purpose, but the limbs of a human or a vertebrate animal, in which the compressive forces are counteracted by rigid bones, are capable of more rapid, precise movements. For a given function such limbs are also smaller in cross section. Considerations of this kind help to explain the anatomy of flying animals such as birds and bats: not only the structure of their wings, but their legs and feet, which are rigid enough to enable them to perch on trees, and to act as weapons; yet these structures are thin enough to have a minimum aerodynamic drag.

As we previously pointed out, Nature never produces an all-rigid animal like an engineer's machine. The tension loads are nearly always carried by tendons, muscles, and membranes. Plants, on the other hand, do not have to be mobile. However, the whole structure of small plants and the growing shoots and leaves of large ones are vulnerable to mechanical damage and are most safely made from soft tissues which, in any case, are best suited to the growth processes. Considered from the point of view of evolution, the competitive requirements for plant structures have been rather different from those governing the development of animals. Plants do not need sharp teeth or bones permitting quick movement, but they do need to be able to compete for the available sunlight, and sometimes for moisture.

These competitive arenas have included immense prairies covered with short grasses, but more often the evolutionary battlefield has been the forest or the jungle. Here victory has tended to go to the tallest plants. Tall, stiff varieties of grass—bamboos—sometimes shot up in a single season to heights of 60 feet (18 meters). In the long run, though, the slower-growing, longer-lived, and still taller plants, the trees, have usually been the winners. Trees have been so enormously successful and wood has become so familiar that we do not often appreciate what a sophisticated and ingenious material timber is.

Bamboo is a tall, stiff grass. It may reach heights of 60 feet in a season.

An ingenious and sophisticated material—wood—photographed at Saint-Cloud by Eugène Atget in 1922.

WOOD: THE STRANGER IN OUR MIDST

The central character of G. K. Chesterton's Father Brown stories is a little priest-detective who specialized in solving crime problems by his respect for the familiar and the commonplace. He would succeed by recognizing the blindingly obvious solution, solutions so obvious that they had been ignored or despised by the clever professional experts. He would probably have been very much alive to the structural virtues of wood.

To begin with, in terms of sheer quantity, wood is undoubtedly the most plentiful purposive structural material in the world. The tonnage of timber existing in living trees at this moment is beyond computation, but it is certainly of the order of many billions of tons. If we consider the quantity of timber felled in sufficiently industrialized contexts to find its way into the official statistics, the annual world consumption is in the region of 1,000 million (10^9) tons. This does not include rough timber used for fuel or

The de Havilland Mosquito bomber, which was built of wood, was one of the most successful aircraft of World War II. Some 7,781 were built. In addition, about 5,000 wooden gliders were built in England for the invasion of Normandy in 1944.

Recall that a structure loading coefficient is the ratio of the load borne by a system to the distance it must be carried within the system.

for primitive housing, fencing, and similar purposes. The world's consumption of metals is around 500 million tons per year. Contemporary fears of a shortage of timber are greatly exaggerated. It is true that some forests, notably the rain-forests of Brazil, have been destroyed or depleted, mostly by bad management. But in the Pacific NorthWest, the replanting policy of recent years seems to be effective and one has only to drive about in, say, Vermont to feel that the trees are closing in on one. This is true in many other parts of the world. If there is ever a real shortage of timber it will represent a triumph of mismanagement.

Timber can be said to have an "image problem" in that not only is it very plentiful and familiar, but it is used in an extraordinarily wide spectrum of technologies, many of which are not "advanced." Traditionally timber has always been used for constructions like log huts and farm fencing, which do not require great skill to achieve. Wood has also been used for thousands of years in crafts, such as carpentry and boatbuilding, which do demand higher manual skills but which have not until recently accommodated scientific ideas. In modern mechanized furniture factories, the emphasis has usually been on low production costs and attractive appearance rather than on structural engineering design. Finally, much wood finds its way into the paper and board mills—which, incidentally, operate by means of very effective and advanced technologies of which the majority of people are ignorant.

CHAPTER SEVEN

These manifold applications do not encourage the widespread appreciation of wood as an efficient and sophisticated structural material. At best, wood is apt to be considered as a material suited to the revival of "craftsmanship." But this view does not take into account the influence of structure loading coefficients (Chapter 3) upon the choice of materials. Where the structure loading coefficients are high, as they generally are in machinery—metals, such as steel, are generally preferable. Moreover, metals are hard and temperature-resistant: wood is not a suitable material for crankshafts. But where the structure loading coefficients are low, as they are in trees, boats, and aircraft, wood is a very efficient material indeed.

The nineteenth-century way of thinking that engineering glamor and virtue are associated with highly loaded structures has persisted to the present day. Materials that are good for making engines are deemed to be good for making airframes. But the structural and material requirements of the two applications are quite different and cannot, as a rule, be met efficiently by a single material such as steel.

Too much publicity and too much of engineering education has been concentrated on highly loaded structures and on the materials that are suitable for making them. In fact, the large majority of buildings, containers, airframes, and other technological structures are quite lightly loaded in relation to their dimensions, and are generally much more expensive and much heavier than they really need to be. In spite of their vast economic importance, only recently have the scientific problems associated with lightly loaded structures begun to be taken seriously.

Almost without exception, biological engineering produces lightly loaded structures. As a consequence, wood may be regarded as the supreme structural material in biology. In the higher technologies wood has shown itself to be an excellent material for aircraft construction. It is much in use today, mostly in molded and laminated forms, for the construction of yachts. In the current extensive American wind-turbine program for the generation of electricity, wood is used for the blades of the enormous windmills.

Steambent beechwood armchair, Gebrüder Thonet, around 1900.

THE CONSTITUTION OF WOOD

The most important structural constituent of wood is cellulose. In chemical terminology *-ose* denotes a sugar, so *cellulose* is the sugar associated with plant cells, or rather with cell walls. Cellulose is in fact a high-molecular-weight, long, linear polymer of the simple common sugar, glucose.

Glucose is synthesized in the leaves of trees and other plants by the action of sunlight on water and carbon dioxide, with the aid of the green catalyst chlorophyll. The glucose molecule is a ring structure that is fringed with hydroxyl (-OH) groups. Because of these hydroxyl groups and because of its relatively low molecular weight, glucose is water-soluble. It is thus conveyed in solution in the sap through the various cells and vessels of the plant to the growth zones. In the growing cell wall, other catalysts, notably the auxins, take charge, and the glucose molecules are polymerized or joined end-to-end by a condensation reaction that produces the linear long-chain cellulose molecule.

The long cellulose chains lie roughly parallel to one another in the cell wall. When they get the opportunity, the cellulose molecules crystallize into parallel bundles known as *micelles* or *crystallites*. Usually, crystals are much larger than the molecules from which they are made, but in the case of cellulose, the crystallites or micelles are usually shorter than their molecules.

Because of its high molecular weight, cellulose is not water-soluble in the ordinary sense, although it can and does absorb water molecules around its free hydroxyl groups. The crystallites, however, are not accessible to water. Much of the virtue of natural cellulose is due to the presence of these crystallites, which are strong, stiff, and impermeable to water. Celluloses that have been chemically dissolved and artificially reconstituted, such as cellophane are much less crystalline, much more affected by moisture, and also mechanically inferior to natural forms of cellulose such as wood, cotton, and flax.

Left, cellulose. Right, glucose molecule.

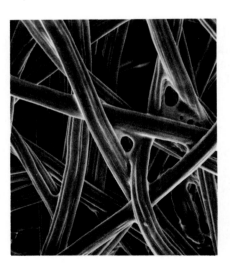 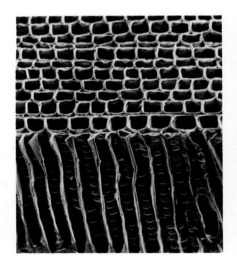

Natural cellulose in the solid state has a specific gravity around 1.5, much like other sugars. It is, of course, a good deal lighter than the engineering metals. Weight for weight, the Young's modulus, E, of cellulose is nearly the same as that of the metals, and its tensile strength is better. However, as we mentioned in Chapter 3, trees are structures that are fairly lightly loaded in relation to their size, and the loads they support are mostly compressive and bending ones. In such cases it does not pay to use a "solid" structure—that is, one whose sections contain no holes or gaps. For comparable loading conditions, engineers tend to use their metals in the form of tubes. Nature uses this strategy (among others) in grasses such as bamboo and, of course, in bones.

However, grasses are fairly ephemeral structures that shoot up in a single season by a process described in engineering terms as *extrusion*. Trees often take hundreds, sometimes thousands, of years to grow; each year the growth process lays down a thin new layer of woody material between the bark and the older wood. After a tree has been felled, these growth layers or annual rings are usually clearly visible, and by counting them, the age of the tree is easily ascertained.

The nature of the arboreal growth process prevents the tree trunk from taking the form of a hollow tube like bamboo. In fact, macroscopically, the tree trunk is nearly (though not quite) uniform from its center out to where the bark begins. Thus, regarded as an engineering structure, the tree must be considered as a solid cylindrical beam or column. For such members the criterion of structural efficiency is $\sqrt{E}/\rho$ (Chapter 3), where E is Young's modulus (a measure of stiffness) and ρ is the density. In other words, it pays to reduce the density. A practical way of reducing the density of a material like wood is by cellularization—that is by putting holes full of air in it so that we end up with a bundle of more or less empty tubes resembling a honeycomb. It is true that by doing so we reduce the absolute value of the Young's modulus, E. But in wood this reduction in E is closely proportional to the reduction in density. Consequently, the denominator ρ decreases faster than the numerator $\sqrt{E}$, because as the denominator decreases by a factor of x, the numerator decreases by a factor of only $\sqrt{x}$. Hence the overall ratio—and the efficiency of the structure—increases.

The hollow-tube cross section of bamboo.

Here we come to a sort of teleological difficulty in biomechanics. In engineering design, most devices have a single purpose. In biology, devices often have multiple purposes. A device that originally functions to facilitate growth may end up by serving some quite different need. In plants, and especially in trees, the hollow tubular cells begin as a growth mechanism, because their cavities act as pipes to convey the necessary sap up and down the tree, a distance that may be as great as 300 or 400 feet. Thus in the *sapwood*, the newly formed outer layers of the tree close to the bark, the cell cavities are largely filled with sap, which is mostly water. As the trunk gets thicker, the earlier cell layers get covered by more recent growth and come to form part of the *heartwood*, the core of the tree. Heartwood is biologically inactive, and its role is now a structural one; much of the water in the cell cavities is replaced by air so that there is a large reduction in density. In the grown tree, structural efficiency is what the cellular morphology is "for."

Although in virtually all trees the actual specific gravity of the cell wall material (which is mostly cellulose) is much the same, around 1.4 or so, the effective density of the timber varies greatly—from about 0.15 for balsa up to nearly 1.0 for lignum vitae—a very hard, heavy wood used in making pulleys and gears. These, however, are the extreme values. The most successful species botanically—and the most useful technologically—have specific gravities around 0.4 to 0.6. Sitka spruce, which is favored for aircraft construction, has a specific gravity of about 0.38 in its seasoned condition. California redwood (*Sequoia sempervirens*), which at a mature height of about 360 feet is by far the largest living thing, is slightly more dense, around 0.41. Hardwoods such as oak and ash have densities of about 0.6 when they are dried.

The densities and Young's moduli of the better kinds of wood result in values of $\sqrt{E}/\rho$ that are around five times as high as those for steel. Hence wood is five times more efficient than steel for struts and columns. Nature does not construct panels from wood but carpenters and engineers do. For this purpose the efficiency criterion is $\sqrt[3]{E}/\rho$, which gives for softwoods a value about eight times that of steel (see the table on page 51).

Again, Nature does not drive nails and screws into wood, but carpenters do. This practice is made feasible by the cellular structure of wood. When the screw or the nail is driven into the wood, the cell walls on either side of the fastening are deflected elastically and so grip it. Provided that we have used a reasonable amount of skill, only local crushing of the cells will occur, and the wood will not be split.

THE STRENGTH OF WOOD AND ITS MECHANISMS OF FAILURE

Trees are actually vertical cantilevers that may have to stand up to winds that might blow from almost any direction. They must have a symmetrical, roughly circular cross section so that when the tree is bent the resulting compressive stress on one side (the lee side) will have a maximum that is equal and opposite to the maximum of the resulting tensile stress on the other side (the windward side). However, because wood is a cellular material and the cell walls are usually quite thin, wood is much weaker in compression than it is in tension, because the walls of the cells buckle and collapse locally under comparatively small stresses. The compressive strength of wood along the grain is usually only about a third or a quarter of its tensile strength. Thus trees might be expected to break rather easily by bending in strong winds as a result of local compressive failure. This does not happen because trees take several ingenious precautions against this danger.

The tree grows in such a way that the outer layers of the trunk are prestressed in tension at a stress somewhat greater than 2,000 lb/in^2 (27 MN/m^2). Because the compressive strength of wood is around 4,000 lb/in^2, this prestressing improves the effective bending strength of the trunk by 50 percent or so. The corresponding tensile stress on the other side of the trunk is proportionately increased, of course, but generally has plenty of tensile strength in hand.

The prestressing of a tree trunk. Because wood is much weaker in compression than it is in tension, the tree grows with its outer layers prestressed in tension. Thus the solid tree trunks that were often used in traditional technologies (e.g., for ship's masts) were considerably stronger in bending than more sophisticated structures made from sawn timber.

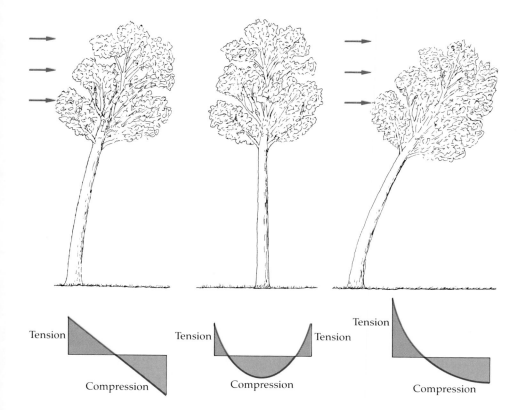

This prestressing of the outer layers of tree trunks justifies the traditional belief of ship's sparmakers and house carpenters that it was deleterious to remove (by adzing, sawing, or planing) the outer layers of a tree that was to be made into a mast or part of a roof. These craftsmen rightly preferred to use the treetrunk as nearly as possible in its natural condition, removing only the bark and the minimum amount of underlying wood. Although modern engineers were often scornful, these craftsmen were generally right. Unfortunately, it is not practicable to use timber in this form in the construction of aircraft or modern yachts.

The prestressing of the outer layers of the tree in tension is analogous, in technology, to the use of inflated or fluid-filled structures such as tires, balloons, or Dracones. However, prestressing in tension in all-solid systems comparable to trees, seems to be rare in technology. More often, the prestressing system works the other way around. That is, the weak component is put into compression—as it is in prestressed concrete, where the cement, which is weak in tension, is kept more or less in a compressed state by steel rods or wires.

We consider now the mechanisms of compressive failure in wood. Many laymen—and too many engineers—tend to think of mechanical failure in terms of the tensile fracture mechanisms. As we have said several times, failure in compression is at least as dangerous as failure in tension, and the provision of adequate compressive strength undoubtedly accounts for much more structure weight and much more expenditure of money (or metabolic energy) in technological (or biological) structures than does the provision of tensile strength.

The most common form of compressive failure that has to be guarded against is Euler failure, or failure by buckling. As we have seen, by virtue of its low density and relatively high stiffness wood is exceptionally good at resisting buckling failure. However, "them as won't bend, breaks," as the saying goes. If wood does not fail in compression by overall Euler-style elastic buckling, then sooner or later it will fail by a local crushing or creasing of the cell walls.

In many materials local compression failure can be sudden, almost explosive, and very dangerous indeed. Local compression failure in hard, dense, homogeneous materials such as brick, stone, and cement is initiated by shear cracks running at about 45° to the applied compressive stress. Such cracks spread easily, and often the structure ends by being very extensively crushed, with dangerous splinters thrown off at a high velocity. Usually there is no possibility that the structure will remain serviceable after its initial failure.

In wood, although the stress at which local compression failure begins is fairly low, the process is a gradual one that is, on the whole, remarkably

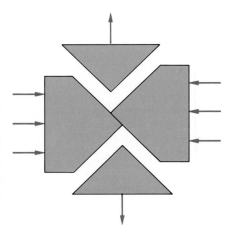

Explosive failure in compression by shearing in a solid, homogeneous material.

safe. Because of the hollow cellular morphology of wood, the result of failure is not really a crack but a crease caused by the local buckling of the cell walls. Moreover, the plane of the crease usually lies not at 45°, but roughly perpendicular to the applied stress. In other words, in a tree trunk or a wooden beam the creases generally run straight in toward the center. Thus, as the material fails, it tends to close up on itself rather than break in two. Also, because the creases toward the middle of the tree trunk or beam, they approach the neutral axis, where, of course, the stress is nearing zero. This is why compression creases in wood, and especially in wooden beams, tend to proceed a short distance in from the surface and then stop.

Compression creases in wood are often a stable condition. If the wood is compressed still further, rather than extending the existing crease deeper into the wood the additional load creates new creases. Although in force terms the stress needed to initiate a compression crease in wood is usually quite low, which is why wood is regarded as weak in compression, in energy terms the energy needed to propagate the crease is very high indeed. It is many times higher than the energy needed to propagate a tension crack, even in a tough material. Thus when we apply the ideas of modern fracture mechanics to wood in compression we find that complete and catastrophic failure is unlikely, which is one of the reasons why wood is such a safe material.

Having said this, we must now insert a caveat. Although these creases are unlikely to spread and cause failure in compression, they have very little strength in tension. Thus if the load is reversed, there may be a sudden failure. Wooden beams such as oars, which are highly and frequently stressed in one direction, often accumulate extensive collections of little creases along their compression faces. These are safe enough as long as the oar is loaded only in the usual way. If the direction of bending is reversed for some reason the oar may break suddenly and unexpectedly at a surprisingly low load.

This sort of failure would not usually matter much with oars, but fatal cases have occurred with wooden aircraft such as gliders. These aircraft had been subjected to heavy, clumsy landings which had caused compression creases in what was normally the tension boom of the wing spars. When the gliders were later pulling out of a dive, the wings would fall off.

As we mentioned, usually the mechanisms by which the cell walls in natural wood buckle under compression are such that the resulting crease runs approximately at right angles to the compressive stress; this, as we have been saying, is normally a safe condition. However, in reconstituted wood products and in artificial fiber-reinforced composites the structure of the material is more or less solid rather than cellular. Thus there are no

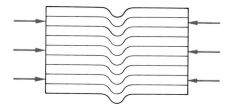

Compression failure in wood. When wood fails in compression, the failure generally takes the form of a crease that is perpendicular to the applied load. This is a very "safe" mode of failure.

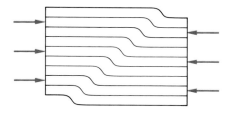

In reconstituted wood products and in artificial fiber composites compression failure is apt to take the form of a crease running at 45°. This is a more dangerous mode of failure than the right-angled creases in natural wood.

In a heavy landing (top), the wing tips of a glider are forced downward due to inertia forces. In subsequently pulling out of a dive (bottom), the wing tips are forced upward. The weakened lower or tension spar booms are now in tension and may not be strong enough to endure.

holes into which the fibers can buckle. As a result, a compression crease forms at about 45° to the stress, and it is liable to spread in a sudden and dangerous way.

We have seen why cellular wood is fairly safe in compression. We turn now to the question of why wood seldom breaks in tension. First of all, because the morphology of timber enables it to develop a high proportion of the strength of the bonds in the cellulose molecule, the actual tensile stress at which it does break is very high. Usually the breaking stress is about 15,000 to 20,000 p.s.i. (100 to 140 MN/m^2). Since the density of wood is around one twentieth that of steel, the tensile strength of wood is equivalent, weight for weight, to a steel having a tensile strength of about 400,000 p.s.i. Such steels exist only as very brittle wires. Even the "high-tensile" steels of commerce seldom have tensile strengths much above 250,000 p.s.i. and they are relatively brittle.

However, wood is not only strong in tension, it is also very tough, with a work of fracture of at least 1.0×10^4 J/m^2. Weight for weight, this would be equivalent to a steel with a work of fracture of about 20×10^4 J/m^2, which is about the actual value for ordinary commercial mild steels with tensile strengths around 60,000 p.s.i. (400 MN/m^2). In other words, wood is a material with a tensile strength as high as that of the strongest steel (which is very brittle) and a toughness equivalent to that of the toughest and most ductile steel (which is quite weak). We clearly have a great deal to learn from wood.

There is no great mystery about the source of wood's tensile strength, but even as recently as the early 1970's the high work of fracture could not be explained by any of the known work-of-fracture mechanisms. About this time Dr. Giorgio Jeronimidis came from Italy to work with me at Reading University in England. Because the adhesion between the cells in wood is fairly weak—as seen by the ease with which wood splits along the grain—wood is very susceptible to the Cook-Gordon mechanism (Chapter 4): the crack-stopping effect induced by the poor lateral adhesion between cellulose molecules. The resulting jagged, fibrous failure reflects the toughness of wood. However, when Jeronimidis and I had done the sums based on this sort of fracture, the calculated works of fracture were still an order of magnitude short of the measured experimental values.

The explanation had been waiting in the wings. Around 1966 the British botanist R. D. Preston pointed out that the morphology of cellulose fibrils in the cell walls was unexpected. In most cases the fibrils did not lie parallel to the axis of the cell but formed a steep helix at an angle between 5 and 20° to the grain direction. Moreover, in any one tree these helices all had the same sign (all wound the same way) and about the same angle. However, at the time nobody seems to have associated this morphology with the structural virtues of timber, probably because the science of fracture mechanics was not widely understood.

George Jeronimidis was able to show that when a tubular helical structure such as a wood cell was loaded in tension, it behaved in a straightforward Hookean manner up to a strain of about 2.0 percent. In such an arrangement the helical fibers are contracting toward the axis of the tube. That is, they are compressing the cell radially and circumferentially and, sooner or later, the thin cell wall will buckle inward and collapse.

Collapse does occur in wood at around 2 percent tensile strain. The failure is marked by a sharp yield in the elastic behavior followed by a long region of plastic deformation. This been confirmed both by observation of the wood itself and by testing models. The result, for a single wood cell, is startlingly like the elastic-plastic stress-strain curve for a ductile metal such as mild steel. Thus the helical mechanism accounts elegantly for the high work of fracture of natural wood. Also, as we shall see in Chapter 8, it provides a model for the construction of improved synthetic fiber-composite materials.

In materials other than wood, work-of-fracture mechanisms usually depend, in one way or another, on breaking and re-forming intermolecular bonds. Because chemical bonds become less mobile as the temperature falls, solids tend to become brittle at low temperatures, and this can be dangerous. But the helical work-of-fracture mechanism does not depend primarily on breaking and reforming bonds, it depends rather on a form of elastic buckling. For this reason Jeronimidis found that the work of fracture of wood *increases* at low temperatures. Since many trees grow in cold climates where the worst storms are likely to occur in winter, this is a considerable advantage. It is also, of course, an advantage for wooden aircraft, sleighs, and skis.

SWELLING AND ROT

Although the crystalline part of cellulose is inaccessible to water, about 50 percent of the cellulose fiber is not crystalline. The cellulose molecules are provided with many hydroxyl groups which attract each other (and thus hold the fiber together) and also attract water. As they become surrounded by water molecules, the mutual attraction of neighboring hydroxyl groups is weakened and the cellulose swells laterally.

The moisture content of wood depends upon the relative humidity of the surrounding air. In a dry climate, such as that of Nevada, the equilibrium moisture content of wood may be as low as 5 percent. At 95 percent relative humidity the moisture content will be around 25 percent. Up to

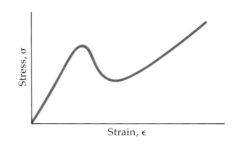

The stress-strain curve for a helical, tubular wood cell in tension. The similarity to the S-curve behavior of mild steel is remarkable.

this value all of the moisture in the wood is held in the cell walls, and is associated with the hydroxyls in the cellulose. Beyond a level of 25 percent, the moisture in wood takes the form of loose water in the cell cavities.

These changes in moisture content cause weight changes that are of minor consequence in technology; it is the resulting dimensional changes that are important. Although most species of timber change little in length with changes in moisture content the lateral movements are considerable. For every 1.0 percent change in moisture content, wood will shrink or swell laterally by about 0.5 percent. Since moisture content changes of up to about 20 percent are quite likely to occur if the environment of the wood is changed, the resulting dimensional swelling or shrinkage may be as high as 10 percent—which, of course, can be a very serious matter.

In traditional wooden structures—log huts, shingled roofs, old-fashioned ships, and so on—allowance has been made for shrinkage and swelling so that these effects do not matter very much. However, when modern engineers come to put together "up-to-date" structures—especially aircraft—using rigid fastenings and adhesives, there can be very serious trouble. If timber is restrained from shrinking when it wants to do so, it is very likely to split. If it is restrained from swelling when it picks up moisture, the resulting forces can be very large indeed. Failure to make allowance for these effects was the cause of many fatal accidents to wooden aircraft during World War II.

As we have said, traditional wooden structures are usually sufficiently flexible, in one way or another, to accommodate shrinkage and swelling. Often modern sophisticated structures, in which the wood is used in thin sections such as veneers, which are glued together with modern adhesives, are also free from trouble. But these practices require highly educated and experienced designers, such as are to be found in the best yacht-building firms. What is to be avoided is the crude imposition of "modern" ideas upon a very subtle traditional material.

Swelling and shrinkage due to moisture content are not the only specifically biological drawbacks of wood. Another of its behaviors may constitute an even greater hazard: its vulnerability to rot.

The process of evolution depends on death. After death the body must decay so that the products of decomposition can feed other forms of life. If trees did not rot, the world would be cluttered with dead wood and the cyclic system of Nature would be frustrated. The management of rot in timber has shaped both technological and human history—especially in North America.

Timber rot is caused by fungi that live parasitically on cellulose because they have no chlorophyll, so they cannot photosynthesize sugars for themselves. The spores of various fungi are nearly always present in wood-

"I play for Seasons; not Eternities!" says Nature.

George Meredith (1828–1909)
"Modern Love"

Rot and fungi. Polystictus fungi on a birch log.

work, but they do not become active unless the conditions are favorable. Rots cannot flourish if the moisture content of the wood is below about 18 percent. If timber is kept reasonably dry and properly ventilated, it is unlikely to rot. However, much depends on the climate. In Britain and in much of northwestern Europe, timber used outdoors is likely to have an average moisture content of 20 to 25 percent. It is therefore liable to decay unless it is regularly painted and maintained. For this reason—apart from the relative shortage of growing timber—wooden structures have tended to be replaced by masonry or metal. There are few wooden houses in England, and almost no wooden bridges.

The Union Pacific's Engine Number 119 crosses the Promontory Trestle in Utah on its way to the linkup with the Central Pacific, 10 May 1869. Trestles and other large wooden structures (like the covered bridge on page 127) are common in North America, where the climate is drier than in northern Europe.

In North America, however, the climate is such that wood tends to settle down to a considerably lower moisture content, often around 10 to 12 percent. This is well below the critical level for rot. Wood, especially softwood, was and is much more plentiful in many parts of North America than in Europe, and the pioneers acquired legendary skill with their axes. The low likelihood of rotting combined with the abundance of wood ensured that wood was used to construct many houses, trestle bridges, and so on—structures that even if they were unpainted and neglected lasted much longer than they would have in Europe.

Wooden structures were generally cheaper than the equivalent brick, stone, or metal ones, and they could often be put up more quickly, and be altered more easily. The expansion of the North American railroad system was greatly facilitated by the extensive use of wooden trestle bridges. In fact the cost of the nineteenth-century American railroads, per mile, was about a fifth of that of British railroads, although American wages were perhaps twice as high as those prevailing in Britain at the time. The American wooden bridges seem to have needed very little maintenance. Many of them are still in existence, being now well over 100 years old. In contrast, the large wooden railroad viaducts built in Cornwall England in Victorian times decayed fairly quickly and were replaced long ago by steel and masonry.

BONE

As another example of material design, bone forms an interesting comparison with wood. Bone starts life in the embryo as collagen, a protein which, as we have seen, is used structurally in tendon and as a reinforcing fiber in

The "skeleton," as we see it in a museum, is a poor and even a misleading picture of mechanical efficiency. From the engineer's point of view, it is a diagram showing all the compression lines, but by no means all the tension lines of the construction; it shows all the struts, but few of the ties, and perhaps we might say *none* of the principal ones; it falls to pieces unless we clamp it together, as best we can, in a more or less clumsy and immobilized way. But in life, that fabric of struts is surrounded and interwoven with a complicated system of ties—"its living mantles jointed strong, With glistering band and silvery thong" (Oliver Wendell Holmes): ligament and membrane, muscle and tendon, run between bone and bone; and the beauty and strength lie not in one part or another but in the harmonious concatenation which all the parts, soft and hard, rigid and flexible, tension-bearing and pressure-bearing, make up together.

Sir D'Arcy Thompson
On Growth and Form (1917)

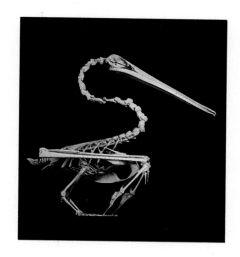

other soft tissues. Collagen is tough and it has a J-shaped stress-strain curve (page 148). It has excellent tensile properties, but, being flexible, it is not fitted to carry compressive or bending loads, which, of course, is what bones are chiefly for. Moreover, in the economics of biology, collagen is metabolically expensive.

Early on in the growth process, therefore, the animal starts adding to the collagen thin stiff fibers of the inorganic calcium compound hydroxyapatite [$Ca_5(PO_4)_3(OH)$]. Hydroxyapatite is very much stiffer than collagen, and considerably cheaper in metabolic terms. Most adult bones end up with a hydroxyapatite content of around 67 percent.

In bone the structural unit that is analogous to the plant cell in wood is the *osteon*. Like the plant cell, the osteon is a hollow tube whose cavity also serves to convey a liquid—in this case, blood. Again, like the plant cell, the walls of the osteon are constructed of a helical arrangement of thin fibers; in bone these are fibers of collagen alternating with parallel fibers of hydroxyapatite. In bulk hydroxyapatite is not particularly strong, but in the form in which it exists in bone, the fibers are both smooth and thin. They are, in effect, equivalent to whisker crystals, which, as we saw in Chapter 4, are often very strong indeed.

The resulting material is, as a short column, considerably stronger in compression than wood and a little stronger in tension. However, when the comparison is made on a weight-for-weight basis, bone is less impressive. As a long column in compression, bone does not begin to compare with wood. When Nature elects to produce in an animal, such as a giraffe,

Photomicrograph of fractured bone.

Approximate mechanical properties of wood, bone, and antler

Material	Specific gravity	Work of fracture (J/2)	Tensile strength		Compressive strength		Young's modulus	
			p.s.i.	MN/2	p.s.i.	MN/2	p.s.i. ×10⁶	MN/2 ×10⁴
Wood (pine)	0.4	15,000	15,000	100	4,000	27	1.6	11
Bone	2.0	2,000	18,000	120	22,000	150	2.7	18
Antler	1.9	15,000	29,000	200	33,000	230	2.0	14

Approximate weight-for-weight properties of wood, bone, and antler in comparison with mild steel

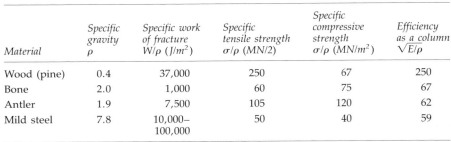

Material	Specific gravity ρ	Specific work of fracture W/ρ (J/m²)	Specific tensile strength σ/ρ (MN/2)	Specific compressive strength σ/ρ (MN/m²)	Efficiency as a column $\sqrt{E}/\rho$
Wood (pine)	0.4	37,000	250	67	250
Bone	2.0	1,000	60	75	67
Antler	1.9	7,500	105	120	62
Mild steel	7.8	10,000–100,000	50	40	59

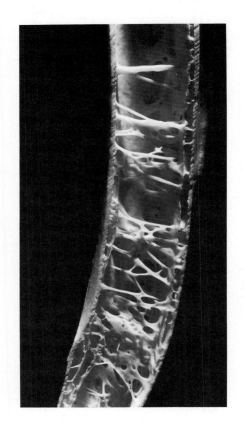

Cross section of an eagle bone. The arrangement of supporting structures inside the bone recalls Sir D'Arcy Thompson's observation that a vulture wing resembles a Warren truss (see Chapter 3).

the equivalent of a long Euler column, she appears to rely upon active rather than passive elasticity to keep the column upright. That is, the corrective action of muscles and tendons is what prevents buckling. So perhaps the rather low value of $\sqrt{E}/\rho$ for bone is not much of a handicap in practice.

An apparently more significant disadvantage for bone as a structural material is its low work of fracture, which is something like a tenth of that of wood. One reason for this is that the walls of the osteon are much thicker than the cell walls in wood. In fact, they are much too thick to buckle in the way that cellulose walls do. This means that the ductility mechanisms Jeronimidis found in wood cannot work in bone.

The work of fracture that bone does possess is almost entirely due to the Cook-Gordon mechanism (Chapter 4): that is, to the crack-stopping effect of the poor lateral adhesion between the osteons. In fact, the measured work of fracture of bone is pretty close to the value that Jeronimidis and the author calculated as likely for wood, supposing that the helical collapse mechanism did not exist. As we said in Chapter 4, secondary cracks in healthy bone are fairly common, especially in old people. Only under abnormal conditions, such as those appertaining to deep-sea divers (page 100) do they have any serious effect.

The limited work of fracture of bone has a scale effect upon the safety of animal structures (see tables on facing page). The structure and the mechanical properties of most of the bones of most vertebrate animals, both large and small, are very much the same. In fact, bone comes near to being a standard material, rather like an engineering metal; this similarity extends to its work of fracture.

Thus the critical Griffith crack length (Chapter 4) for bone is more or less constant, and lies somewhere around 3 centimeters. This is true both for mice and for elephants—and the consequences for elephants are serious. Small animals with bones less than a couple of centimeters thick generally have no very serious fracture mechanics problem. Mice, cats, dogs, and monkeys seldom break their bones although they are often very active. Animals of moderate size, such as humans and horses, are somewhat more prone to bone fractures. Horses, which are larger than humans, seem to be forever breaking their legs. Athletes and skiers do this rather less frequently. Our skulls, which are comparatively large shell structures, are especially vulnerable—a very good reason for wearing crash helmets. Large animals like elephants, whose main bones are much more than 3 centimeters thick, have to behave in a fairly sedate manner—which is what they all seem to do. This is an aspect of the extinction of still larger animals, like dinosaurs, which I have not seen commented on in the biological literature.

An analogous state of affairs exists in engineering at the present time. Because small and medium-sized metal structures have proved to be safe in service, engineers and naval architects have scaled them up, assuming that if the stresses remained the same the large structure would be as safe as the smaller one. But the critical crack length in the mild steel plates used in engineering is between 1 and 2 meters. The history of accidents to large ships, such as supertankers, during recent years has been unfortunate, and perhaps engineers might look with profit at the extinction of dinosaurs and at the sedate conduct—and the scarcity—of elephants.

Although the trunks of trees are often very much thicker than the bones of animals, the critical crack length for wood is around 50 centimeters, which is 15 or 20 times that of bone. Of course, unlike wood, the bones of most animals are more or less protected by fairly thick cushions of soft flesh, fur, and so on. This may be one reason why Nature does not seem to have made any serious attempt to develop bones with a higher work of fracture. However, this excuse cannot be made for supertankers.

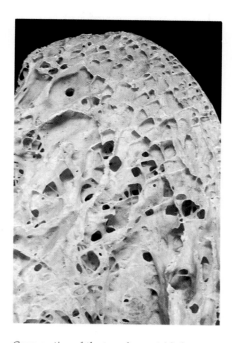

Cross section of the top of an ostrich femur shows that even the largest and heaviest bones have many hollow spaces (for lightness) and many supporting structures (for strength).

ANTLERS: HOW TO BECOME MONARCH OF THE GLEN

Each spring in the Scottish Highlands, the local stags fight each other to establish which of them shall become the "monarch of the glen" and the husband of all the local hinds. The losers have to spend a discontented and celibate summer sulking somewhere out of sight. These fights are not ritual tourneys or he-stag bluff; they are very serious battles, even though the losers seldom seem to get killed or seriously injured. These large and heavy animals charge each other at full speed, meeting head-on with a clash of antlers that can be heard from a considerable distance.

The Monarch of the Glen *by Sir Edwin Landseer (1802–1873).*

As can be seen from Landseer's famous and much reproduced painting, the antlers are many feet across and have numerous pointed branches. One would expect antlers to have excellent strength and toughness, for they provide a textbook case of natural or sexual selection. In fact, because of their shape and their excellent mechanical properties, antlers were widely used by early humans to make pickaxes and other tools and weapons. As can be seen the tables on page 176, the strength and toughness of antler is much better than that of bone, although antler would not be so good for making a column; but then, that is not its function.

The constitution of antler is not greatly different from that of bone. A matrix of collagen is reinforced with fine fibers of hydroxyapatite. The mineral content of antlers is lower, being about 55 percent as against about 67 percent for bone. Unlike bone, the antlers are shed and replaced every year.

Partly for their biochemical interest and partly because of the lessons they might afford for the design of artificial composites, antlers are being investigated at Reading University at present by George Jeronimidis and Mike Watkins. The results so far are interesting, although Jeronimidis and Watkins have not found any clear-cut special mechanism, comparable to that in wood, to account for antler's very high work of fracture, which is nearly an order of magnitude greater than that of bone.

The toughness of antler seems to be due to a detailed and sophisticated control over the fine fiber structure of the material, and especially of the interfacial adhesion. The fracture surface in antler shows much more extensive fiber pull-out than is the case with bone: the whole appearance of the fracture is "hairy." Why Nature has not chosen to make use of this morphology in ordinary bone is not clear. Possibly it would interfere with the self-healing ability of bone: antler is a throwaway material that is discarded each year. Such a fiber system might also be incompatible with some of the metabolic functions of bone.

8

THE NEW ARTIFICIAL MATERIALS

Polymers, Composites, and the Future

Conception, my boy, fundamental brainwork,
is what makes the difference in all art.

DANTE GABRIEL ROSSETTI (1828–1882)

As we have said throughout this book, most of the major difficulties and most of the major improvements in the science of materials and structures have been conceptual in nature. The barriers have existed more in people's minds than in the facts of science. In this respect the Industrial Revolution was, perhaps, too successful. It brought about such sweeping changes in traditional modes of manufacturing and construction that complacency set in. As the Greeks knew, few things can be as dangerous in the long run— or as obstructive to progress—as too complete success.

Before the Industrial Revolution, technology was based on a wide range of materials: on various kinds of wood, stone, and leather, and on rope and cloth made from various natural fibers. Metals such as iron and copper were used too, but only to a moderate extent. The craftsmen who fabricated these materials often had little formal education, and they worked and thought in traditional "unscientific" ways. However, the nature of their trades forced upon them a certain humility toward the physical substances they sought to master.

During the nineteenth century much of this traditional technological culture was swept away, and most of what remained was disparaged as out of date. Wooden sailing ships, the products of thousands of years of evolution, were contemptuously dubbed *windjammers*. Engineers saw their technology as a brave new world of steel and concrete, steam and internal combustion engines. New metal alloys were waiting to be discovered and exploited. The apparently limitless possibilities of the new materials and new mechanical devices lulled engineers into thinking that improvements would continue to be made in the design of machinery and structures, but that the conceptual basis of engineering would not change. There would be no need to endure the emotional discomforts of fundamental and drastic rethinking; certainly there would be no call to reevaluate the work of the old-fashioned shipwrights and coachbuilders. The break with the past that characterized the Industrial Revolution was considered indispensable to progress.

Thus most engineers inhabited a world that was intellectually self-contained and self-sufficient. They drew little inspiration, or warning, from other branches of modern science such as medicine and biology. Plants and animals might indeed be successful in the evolutionary struggle, but engineers did not think that they had anything to learn from them as structures.

As with most sweeping revolutions, a counterrevolution of some kind was more or less inevitable although it took a while to develop in an effectual way. The nostalgic nineteenth-century movements for the "revival of craftsmanship," led by writers like John Ruskin (1819–1900) and William Morris (1834–1896), probably served to harden, rather than to liberalize,

the prejudices of conventional engineers, both in Victorian times and afterwards. In the ideology of these movements, "craftsmanship" implied a rejection of scientific analysis and a reversion to subjective traditional skills, in the interests of aesthetics and job satisfaction. Scientific technology was not only associated with ugliness; it was responsible for the dark satanic mills and for the unacceptable face of capitalism.

Claude Monet's Gare Saint-Lazare: Le train de Normandie *(1877) celebrates the new materials of the Industrial Revolution: iron in the supports and frame and glass in the roof of the train shed. They stand in contrast to the traditional materials and form of the wooden freight sheds in the background.*

Only recently has an eclectic approach to the subject of materials and structures combined a respect for the experience of the past with modern scientific and analytical methods. The new approach is a vigilant one that seeks to profit from analogies with other sciences.

THE INVENTION OF MODERN PLASTICS

The break with the Victorian ways of thinking has been facilitated by the gradual introduction of new nonmetallic, load-bearing materials: *plastics*. Although the development of plastics began in a rather small way around the beginning of the twentieth century, their use has expanded steadily. The current view is that continuing innovations in the field are likely to produce some rather drastic changes in both the ideas and the practice of engineers.

Plastics had their origin not in the inadequacies of orthodox structural and mechanical engineering, but in the need for better insulating materials in the young and growing electrical industries. The first important practical application of electricity was in telegraphy. The insulants available in the early days of electricity—gutta-percha (a coating derived from the sap of a Malaysian gum balsam), shellac, waxed paper, glass, and ceramics—were adequate for the existing technology. In fact, they were good enough to enable a worldwide cable and telegraphic system to be established. Toward the end of the nineteenth century, however, electricity came to be used on a substantial scale for lighting and other purposes, and electrical engineers began to get into trouble. This was especially true in the design of dynamos that were "large" (by the standards of the day): these generators were apt to run at much higher temperatures than their designers had intended, often because the iron cores of their field and armature magnets were not sufficiently laminated.

Because the windings of these magnets were generally insulated with gutta-percha or with cloth or paper impregated with shellac varnish— substances with low melting points—the wires were often distorted by the forces acting on them, and the insulant sometimes melted and ran out of the dynamo. Glass and porcelain had much better temperature resistance, but these materials were too brittle to be used in moving machinery, or, indeed, in many other applications. Improvement came with the development of more sophisticated techniques for laminating the magnet cores, but electrical engineers were sometimes driven to incorporate water-cooling devices in the design of their generators.

In 1906 the Belgian-born chemist Leo Baekeland, who had already made a fortune by inventing and exploiting a photographic printing paper, discovered that a reaction between phenol and formaldehyde produced the first truly synthetic resin, which became known as Bakelite. Phenol-formaldehyde resin can exist in three distinct states. The early product of the reaction is the A stage, a thick, sticky, brownish liquid. Further heating results in the B stage, a dark brown, brittle solid or powder that is soluble in various solvents, including alcohol. When the temperature of the B-stage resin is raised to around 150°C, it first melts and then hardens permanently to the C stage, an insoluble but still rather brittle resin or laquer.

Working from London, Baekeland started the Dammard Lacquer Company which originally marketed three grades of varnish: Dammard, Dammarder, and Dammardest. The company has since achieved some fame as the Bakelite Corporation. The original varnishes were, no doubt, simply B-stage phenol-formaldehyde resins dissolved in alcohol. In the electrical industry they could be used as substitutes for shellac, principally for impregnating paper or fabric. When baked, the resulting insulant was more or less heatproof and had adequate mechanical properties for its rather limited purpose.

C-stage Bakelite resin has good electrical insulating properties and is temperature-resistant, but when used by itself, it is very weak and brittle. Baekeland's company overcame this drawback by mixing the B-stage resin with cellulose fibers—originally in the form of wood flour (fine sawdust). The cellulose increased the work of fracture of the resin to a level that was acceptable (though still pretty low) for use in a wide range of small components, such as electric plugs and switches.

In the manufacture of C-stage components, a quantity of B-stage molding powder was put into a heated steel mold or die that was maintained at a temperature of about 150°C; the mold was then mounted in a hydraulic press. When the mold was closed, the powder melted and filled the interstices of the die. After a short time the resin hardened permanently, having reached the C stage. It could then be ejected as a finished, hard molding.

It is said that the first commercial molding of this type was a mechanical rather than an electrical component: a gearshift knob for a Rolls-Royce automobile, produced in 1917. Previously such parts had to be machined by hand from the solid raw material with considerable labor, so the saving in cost was enormous. The Bakelite molding trade soon found other nonelectrical applications such as ashtrays and bottle caps. However, early Bakelite moldings found their most important market in the manufacture of small electrical devices. This was partly because the plastic insulating

Containers made of phenolic from the 1920s: a Belplastic lidded pot and Shellware beakers and games cup. Bakelite is one trade name used for phenolic (which is short for phenol formaldehyde).

material was easily induced to flow around—and to incorporate—metal plugs and inserts.

Bakelite molding powder was too weak and brittle to be used for larger components such as switchboards and other electrical panels. These continued to be made from slate for a good many years, although slate must have been an inconvenient, tricky, and expensive material to work with. Largely to meet this market, the first laminated sheet materials were developed during the 1920s. Cellulose-based paper or fabric was impregnated with B-stage phenol-formaldehyde resin and then pressed between heated flat steel plattens at a high pressure, usually around a ton per square inch. The manufacture of sheets of a useful size—say, 100 inches by 50—therefore necessitated pressing equipment capable of exerting considerable forces—in this case, about 5,000 tons. The production of these panels consequently required both heavy engineering and a sizable capital investment.

Paper-based phenolic sheet materials were rather brittle by modern standards and had a number of other defects as well. The panels of radio sets were often still made from Ebonite, a form of hard rubber, until after 1930. However, the phenolic sheet materials soon found a market in electrical engineering. The dark brown of phenolic resins limited their use for decorative applications. However, urea-formaldehyde resins were invented in 1924 and melamine resins around 1930; both these polymers are colorless and substantially transparent. But using such resins to impregnate the surface layers of sheet materials, bright colors and decorative patterns could be applied. This has led to a revolution in the look of kitchens and bathrooms. Most people are familiar with Formica.

Laminated sheet materials based on woven cotton fabrics impregnated with phenolic resin are more expensive than corresponding paper-based materials, but they are much tougher. The trade name Tufnol dates from the 1930s, when very durable gears and cams were first made from such materials. A further important application was the consequence of another virtue, the ability to run efficiently under water. Tufnol bearings have replaced lignum vitae (hardwood) for tailshaft (propeller) bearings in ships, thereby solving a serious and long-standing problem in marine engineering. Tufnol is also used for switching components in electric power plants, where the mechanical loads may be high.

However, for critical structural applications such as aircraft parts, both fabric- and paper-based phenolic sheet materials had a number of serious defects. For one thing, their strength and stiffness were much inferior to those of light alloys. For another, the phenolic sheets were less easy to shape. A laminated material made initially as a flat sheet could be bent to a moderate extent in one dimension, but not in two dimensions. For such curvatures, the B-stage material had to be molded to shape at high pressure, using very expensive matched steel dies.

An even more serious defect in these cellulose-reinforced materials was their response to moisture. Slight changes of dimension did not matter a great deal in small electrical fittings or in switchboard panels, but when large pieces of material were rigidly fastened to a metal framework, as in an aircraft wing, even small percentage changes could produce dangerous effects. Even with very careful control of the manufacturing process, it is difficult to get the dimensional change between wet and dry conditions much below 1.0 percent—which is equivalent to about an inch of movement in eight feet. During World War II, at the height of the aluminum shortage, a group of British engineers (the author included) covered the trailing edges of the wings of 12 bomber aircraft with paper-based phenolic sheets that were riveted to the underlying metal structures. Six of these aircraft were sent to North Africa: the rest were kept in England through the winter. In the machines that went to North Africa—which, of course, has a very dry climate—the laminated sheet shrank and cracked along the rivet lines. In the planes that spent the winter in England, the material expanded when exposed to rain and wet snow and buckled to an extent that was distressing to see.

Wood shrinks and swells even more than cellulose-reinforced plastics, yet very successful structures have been made from it when appropriate design methods have been used so that the shrinkage and swelling are not constrained by untelligent attachments to metal. Having perhaps such thoughts in mind, a Cambridge engineering don called Norman de Bruyne invented, in 1937, woodlike composite called Gordon-Aerolite. This material (which was not named after the present author) consisted of parallel strands of high-grade flax fiber bonded together with phenolic resin. The mechanical properties were good, but not superior enough to those of wood to justify the cost. And wood was cheaper, more easily obtained, and easier to fabricate. Although it is true that Gordon-Aerolite was, in some ways, a material ahead of its time, the project was brought to an end by the British government in 1941 under the stress of war and of civil service politics.

Although Gordon-Aerolite was regarded in its day as another expensive failure of government-sponsored research, it did play an important part in stimulating long-term thinking about the strategy of materials design, especially by indicating that new and revolutionary materials did not have to be modeled on metals. The many arguments and disagreements which Mark Pryor (Chapter 6) and I had with Dr. de Bruyne certainly served to stretch our minds, however much they may have annoyed that rather difficult person at the time.

GLASS FIBERS AND POLYESTER RESINS

The next step in the evolution of modern composite materials was also stimulated by requirements that were basically electrical: in this case, by the needs of airborne radar. Although things like radio direction-finding loops in aircraft had sometimes been housed in casings made from cellulose-phenolic mouldings, the electrical properties of these materials were unsuited to the frequencies and other conditions of radar. Some of the early radomes—plastic housings for the antennae of aircraft radar—were made from methyl methacrylate, which, as Lucite, Perspex, and so on had been developed in the 1930s for airplane windows. However, most radomes were large structures subject to aerodynamic loads that were by no means negligible, and methyl methacrylate proved to be too brittle and not really strong or stiff enough for the purpose.

The solution lay in the development of fiberglass, a composite in which both the resin and the reinforcing were, in the 1930s, fairly novel. The resins used belong to a class known as *polyesters*. Besides having good electrical properties, polyester resins have the enormous advantage that they can be molded at very low pressures—or, indeed, at no pressure at all (as in a vacuum). Furthermore, they can be set by moderate heating, or else by the addition of a catalyst at room temperature. These properties meant that expensive hydraulic presses and steel molds did not have to be used, and very large moldings such as boats or radomes could be made on simple cement or plaster forms.

As we said in Chapter 4, A. A. Griffith and others showed around 1918 that thin glass fibers are very strong indeed. In his experiments the strength of the fibers was greatly reduced by the slightest abrasion or other accidental damage to their surfaces. However, after Griffith's time, an organic coating was developed to protect fiber surfaces so that glass fibers could be spun and woven like other textile fibers, without losing much of their strength. As a result, fiberglass became available in the form of thread, felt, and woven cloth. Such material could be impregnated with a polyester resin with highly satisfactory results. Radomes made in this way went into service around 1942, and had quite an important influence on the progress of World War II.

Since that war, fiberglass composites have become enormously popular in the manufacture of boat hulls, car bodies, swimming pools, and hundreds of other large, shaped, shell structures. Although the cost of the material is greater than that of an equivalent weight of wood or steel, the cost of fabrication is generally much less. Moreover, fiberglass does not rot or rust, and its maintenance costs are generally very low.

These applications of fiberglass represent an impressive success story, but the material does have its limitations. It is not suited for high-technology structures such as airframes. Although its strength and toughness are good, the stiffness of glass-reinforced materials compares badly, weight for weight, with that of metals like aluminum. In applications like boat hulls, stiffness, though desirable, is not a first-priority requirement. Nor is stiffness a universal necessity for flying objects: birds and other flying creatures do not seem to need conventional torsional stiffness—a high degree of resistance to twisting forces—at all. However, in aerospace we appear to have got ourselves into a position where the weight of most structures depends, before all things, on stiffness, and especially on torsional stiffness.

The specific stiffness of glass itself, and therefore of glass fibers, is very close to the values for the structural metals. However, in a composite, about one-third of the weight is due to the bonding resin, which is fairly flexible. So, even in a simple, unidirectional glass-reinforced material (that is, one having stiffness in a single direction), the specific Young's modulus—the stiffness per unit weight—will be less than the values for metals (see table below). Furthermore, torsional stiffness in shell structures such as wings and fuselages depends on the provision of stiffness in more than one direction. When the fibers are disposed so as to provide multidirectional stiffness, the comparison with metals becomes very unfavorable.

Specific Young's moduli of various solids

	Specific gravity ρ	Young's modulus E		Specific Young's modulus E/ρ ($MN/m^2 \times 10^4$)
		p.s.i. $\times 10^6$	$MN/m^2 \times 10^4$	
Wood (spruce)	0.4	1.1	1.6	2.8
Iron and steel	7.8	21.0	30.0	2.8
Aluminum	2.8	7.3	10.5	2.7
Alumina (Al_2O_3)	4.0	38.0	55.0	9.5
Magnesium	1.8	4.2	6.0	2.3
Magnesia (MgO)	3.6	28.0	41.0	7.8
Boron	2.3	41.0	60.0	18.0
Beryllium	1.8	30.0	44.0	17.0
Beryllia (BeO)	3.0	38.0	55.0	13.0
Silicon	2.4	16.0	23.0	6.6
Silicon nitride (Si_3N_4)	3.2	38.0	55.0	12.0
Silicon carbide (SiC)	3.2	51.0	75.0	16.0
Diamond	3.5	120.0	170.0	34.0
Glass	2.5	7.0	10.0	2.6

Over the years many attempts have been made to increase the stiffness of glass fibers, but these efforts have been only moderately successful. Large increases (around 100 percent) in Young's modulus have been achieved by making use of beryllium oxide, but this compound is highly toxic and, for this reason, usually unacceptable.

ASBESTOS-PHENOLIC MATERIALS

In 1944, during a visit to the United States, I was shown at what was then Wright Field a fiberglass fuselage which had been made there for a small trainer aircraft. Apart from such specialised uses as radomes, this was, as far as I know, the first successful application of composite construction to what are known as primary aircraft structures. Of course this component had been chosen because, in that particular case, the stiffness requirements were fairly moderate. However, I believe that this plastic fuselage showed some saving of weight when compared with the metal equivalent; it also showed adequate strength on the test frame. It was, in fact, very impressive at the time, although the applicability of this form of construction to aircraft was strictly limited by the low stiffness-to-weight ratio of the material.

Using the Wright Field fuselage as an exemplar, I was able, at the end of the war, to persuade the powers-that-be at Farnborough to let me try to develop composite materials and methods of construction for aircraft which would have substantially higher intrinsic stiffness than fiberglass. This was, of course, a good many years before the invention of the

Approximate properties of some fibers and superfibers used in composite materials

Fiber	Specific gravity	Tensile strength		Young's modulus	
		p.s.i. $\times 10^6$	MN/m^2 $\times 10^4$	p.s.i. $\times 10^6$	MN/m^2 $\times 10^4$
Glass	2.5	0.17	0.25	7	10
Asbestos	3.0	0.2	0.3	16	23
Carbon fiber (high-strength)	2.2	0.27	0.4	20	29
Carbon fiber (high-stiffness)	2.2	0.12	0.175	45	66
Boron fiber	2.7	0.3	0.45	37	55
Silicon carbide whisker	3.2	0.33	0.5	51	75
Silicon nitride whisker	3.2	0.33	0.5	37	55
Kevlar 49	1.45	0.27	0.4	13	20

modern, synthetic, high-stiffness superfibers. Searching around among what was then available, I picked on asbestos, which has about twice the specific stiffness of glass. In those days nobody seemed to be aware of, or interested in, the medical hazards associated with asbestos. Certainly the medical department at Farnborough showed no concern with the matter.

There was, in fact, already on the market an asbestos felt material, impregnated with B-stage phenolic resin, which was sold under the name of Durestos. This material was intended to be molded at high pressures—in the region of a ton per square inch—mainly for applications in chemical plants. However, by applying science and low cunning, we were able to mold and harden the material satisfactorily at atmosphere pressure, using a vacuum bag. This meant that we could construct large moldings, such as aircraft wings, using simple and comparatively cheap equipment. Working on these lines, we were able, by 1951, to construct a full-scale delta wing for the E10/47 jet aircraft.

At the time, the chief official interest in the Farnborough method of composite construction was not so much for making conventional manned aircraft as for the mass-production of unconventional, unmanned, long-range weapons. A large-scale project of this nature was well under way around 1954 when major changes in the policy of long-range weapons caused it to be abandoned. As the health risks associated with asbestos began to be appreciated soon afterwards, this method of construction has largely been given up; although it is still sometimes used in rocketry where the high temperature resistance is crucial.

Like most experimental projects the Durestos aircraft project was educational in many ways, some of which were not strictly technical. Over a number of highly critical years the work at Farnborough on composite structures was supported and encouraged by that very enlightened man, the then Director of the Establishment, W. G. A. (Won't Go Astern) Perring (1898–1951). He unfortunately died, at the age of 52, from heart failure, brought on, I suppose, by overwork. Perring had to back up the plastic structures research in the face of bitter and sustained opposition from the engineering pundits in the Structures Department. These people were rejoicing in the demise of wooden aircraft. They knew, with an almost religious certainty, that Providence had intended aircraft to be made out of metals. Any proposal to make them out of these horrible, vulgar, new plastics was not merely unsafe—it was practically blasphemous.

The Structures Department had to approve the strength of all airborne structures before they were allowed to fly. Unavoidably, we had a series of meetings, or rather confrontations, with the Structures Department about the strength requirements, and the testing, of our experimental structures. I shall never forget the looks of what can only be described as burning hatred which I encountered across the conference table. It was obvious that some of these people regarded me, not as a mistaken enthusiast, but as someone actively wicked. The result of all this was that a regulation was imposed which insisted that all airborne plastic structures must be shown to be at least 50 percent stronger than their metal equivalents. This requirement, the notorious 1.5 plastics factor, has now been abolished, but it was in force for many years. (And to think that laymen commonly suppose that the profession of engineering is "not about people"!)

THE NEW SUPERFIBERS

The incentive—and the money—for the development of advanced composites comes mainly from aerospace and defense. For rockets, satellites, missile housings, and other space vehicles, the need is for materials with a higher stiffness and a lower density than the metals used at present in aviation. Such properties cannot be provided by cellulose reinforcement, quite apart from the other defects of cellulose. The only other naturally occurring fiber with high stiffness is asbestos, which, of course, has medical disadvantages; in any case, the mechanical properties of asbestos are not really outstanding.

If we look through a list of elements and compounds, such as those in the table on page 189, it is clear that there are quite a number of substances that possess a very high stiffness for their weight. However, in the forms in which these solids normally exist, they are almost always unacceptably weak and brittle. As we saw in Chapter 4, under certain conditions—especially surface conditions—nearly every solid can be made very strong indeed. The snag is that a material may be very strong under laboratory conditions, but it will usually be very brittle and liable to be drastically weakened by even trivial damage.

The intelligent use of fibers can circumvent these difficulties. For one thing, if the fiber is thin, the critical Griffith crack length that can exist within the body of the fiber must be very short, so its weakening effect will be limited. Secondly, deliberately designed work-of-fracture mechanisms can be introduced into composite systems so that the material as a whole can be made very tough.

The search for new and sophisticated *superfibers* has become an industry in itself. If one looks through the table on page 189, at first sight both beryllium and beryllia (BeO) might seem attractive, but, again, they are both very poisonous and therefore best avoided. Many of the other materials in this table are promising, however, and most of them have been studied in one way or another as potential superfibers during the past few years.

BORON FIBERS, WHISKERS, AND CARBON FIBERS

One of the earlier fiber-forming substances to be exploited was boron. Fortunately, the emotional resistance to advanced structural materials was not nearly as entrenched in the United States as it was in Great Britain. When the American chemist C. P. Talley produced high-stiffness boron fibers around 1958, his achievement was described by a high-ranking general in the U.S. Air Force—possibly with slight exaggeration—as "the greatest technological breakthrough for 3,000 years." Boron fibers have been used extensively for composites with aerospace applications, but their wider usage has been limited by their high cost.

Boron fibers are made by depositing boron onto a heated tungsten core filament. A gaseous mixture of hydrogen and boron trichloride is passed over the hot tungsten wire to produce the following chemical reaction:

$$3H_2 + 2BCl_3 \longrightarrow 6HCl + 2B$$

The reaction works nicely, but tungsten wire is expensive and the process is quite slow: the output is only 150 meters of fiber per hour.

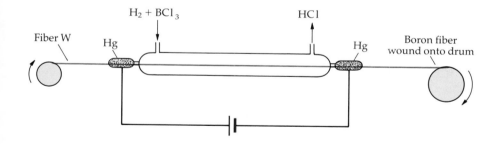

Boron fibers are made by depositing boron from boron trichloride onto heated tungsten wire. The process is, however, comparatively slow and expensive.

Boron fibers are often bonded with normal organic resins, but they are also sometimes used to make reinforced metals. In such cases the boron fiber is usually coated with a surface layer of boron carbide (B_4C) or silicon carbide (SiC) to prevent interaction with the matrix. Neither of these processes is very cheap.

Another type of superfiber has been developed from a process we described the growth of in Chapter 4: very thin, hairlike crystals—whisker crystals—of various elements and compounds. Because the surfaces of

Ceramic whiskers of silicon carbide.

whisker crystals are often smooth, even on a molecular scale, they may show very high strengths. For many years whisker crystals were regarded as scientific curiosities that were unlikely to play any part in serious structural engineering. Nevertheless, beginning around 1956 the author and his colleagues spent a great deal of time trying to grow whiskers of various high-stiffness covalent compounds on a practical scale.

To achieve whiskers of compounds such as silicon nitride (Si_3N_4) or silicon carbide (SiC) we had to use a temperature of 1400°C or more. The metallic element had to be introduced into the reaction in the form of a *transport species,* a compound such as silicon oxide or silicon chloride, which normally exist only at high temperatures.

The furnaces we developed became known as "bran tubs" because we were never sure what would come out of them. We did succeed in making respectable quantities of both silicon nitride and silicon carbide whiskers, which had excellent mechanical properties. Our method was, however, a batch process in which the growth was fairly slow and the cost was high.

Bran tub used to grow ceramic whisker crystals from the vapor phase at high temperature. The process was slow and expensive; newer methods and equipment make for quicker and less costly manufacture of whiskers.

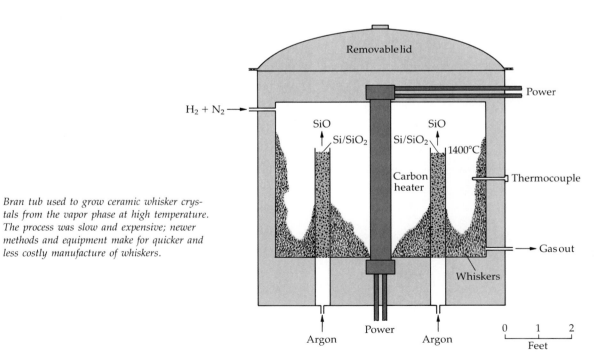

Moreover, these whiskers suffered from competition with carbon fibers, which were developed at Farnborough around the same time.

All the same, high-grade whiskers of silicon nitride, silicon carbide, and alumina (Al_2O_3) have been on the market in kilogram quantities, although they are quite expensive. In the author's judgment it should be possible to grow useful whiskers fairly cheaply, and there may be a breakthrough in this field before long. In fact, silicon carbide whiskers are already being produced in the United States fairly cheaply from rice hulls which provide both carbon and silicon. Because these whiskers are rather short, they are used at present chiefly for reinforcing metals.

We turn finally to carbon fibers which have been represented in recent times as the latest and most glamorous of the superfibers. Actually, in a slightly different form carbon fibers have been with us for well over 100 years. Their first incarnation was as filaments in electric lamps. Britain's pioneer electrical engineer Sir Joseph Swan began to experiment with carbon fibers for lamp filaments around 1860. However, he did not take out his master patent, which described the making of carbon filaments by heating "parchmentized" cotton thread, until 1880. It brought him into conflict with Thomas Edison, who was making lamp filaments by carbonizing fibers of bamboo. Instead of wasting their resources in litigation, Edison and Swan agreed to join forces, and for many years produced "Ediswan" lamps with great success.

Apparently nobody thought at the time to measure the mechanical properties of these carbon filaments, although they may have been quite impressive. The author remembers many years ago, being shown a carbon filament light bulb that had been taken around 1919 from a German U-boat. I was struck by its workmanlike design. Carbon filament lamps had apparently been used deliberately by the Germans in submarines during World War I because they realized that carbon filaments were more resistant to the mechanical shocks of depth charges than were metal ones.

However, by the 1920s carbon filaments were regarded as relics of the past and were forgotten. Not until 1963 was work on carbon fibers—this time for mechanical uses—revived by the British engineer William Watt (1912–1985), who was working in the Materials Department at Farnborough. Like Swan and Edison, Watt began by carbonizing cellulose. However, by some stroke of inspiration he decided to try carbonizing the synthetic polymer polyacrylonitrile.

He heated the original fiber under tension at about 250°C in an inert atmosphere. This produced a ladderlike molecular arrangement. Heating to about 600°C in air oxidized this linked-ring structure to a new form that could be reduced by further heating to a carbon fiber that was, in effect, a chain of graphite rings. Watt discovered that the best tensile strength re-

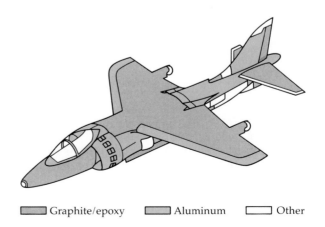

Initial stages of the conversion of polyacrylonitrile into carbon fiber.

Effect of final heat-treatment temperature on the mechanical properties of carbon fibers made from polyacrylonitrile

Final temperature	Specific gravity	Young's modulus		Tensile strength	
		GN/m^2	$p.s.i. \times 10^6$	GN/m^2	$p.s.i. \times 10^4$
1,500°C	2.2	200	29	2.7	40.0
2,600°C	2.2	450	66	1.2	17.5

sulted when the final heating process was halted at a temperature of about 1500°C, but that the highest Young's modulus (that is, the greatest stiffness) was realized by raising the final temperature to about 2600°C.

In the 20 years or so since Watt's pioneering work on the preparation of carbon fibers from polyacrylonitrile, much work has been done, particularly in the United States, to develop cheaper methods of production—notably from pitch fibers. What method will finally be adopted for mass production of synthetic carbon fibers is still uncertain.

Carbon fiber applications got off to a very bad start in England around 1970 with the failure of the carbon fiber fan blades of the Rolls Royce RB211 jet engine. This was due in part to a lack of engineering experience with the material, and in part to unwise political pressure on the firm to make use of a glamorous new material that may have received too much publicity.

Since then, with a more cautious approach and more thorough testing, carbon fiber–epoxy resin composites have begun to be used widely in aerospace applications. The U.S. Navy AV-8B, an attack-type vertical take-off and landing aircraft, already contains 1,317 pounds of carbon-epoxy com-

Graphite-epoxy composites make up 26 percent of the structural weight of the U.S. Navy's AV-8B Attack V/STOL aircraft resulting in a marked increase in payload and overall efficiency.

■ Graphite/epoxy   ■ Aluminum   □ Other

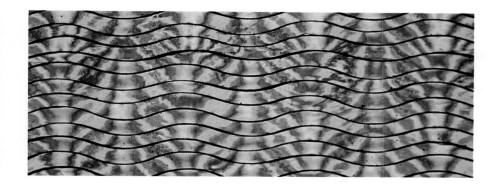

The effect of tensile stress on an experimental composite of undulating graphite fibers in an epoxy resin. Polarized light reveals the distribution of strain within the epoxy matrix. The composite is very flexible at low stresses, but as the tension increases and the fibers straighten out (warping the initially straight edges of the specimen), the material stiffens.

posite in its primary structure, 26 percent of its structure weight. In a projected Grumman aircraft, between 75 and 80 percent of the structure will be built from carbon fiber composites, to save an estimated 26 percent on the all-up weight of this aircraft.

There can be few doubts about the benefits of using carbon fiber composites in high-technology structures; they may well replace metal alloys almost entirely in this field. However, carbon fibers have been marketed for many applications, such as sporting goods and surgical protheses, in which it seems doubtful that so much stiffness is really needed. Such products could probably be made better and more cheaply from fiberglass. In fact, carbon fibers have been represented as a nearly magical material—often by people who are poorly versed in the science of elasticity and who are consequently unaware of the difference between strength and stiffness. For example, promotional copy might talk of the "enormous strength" of carbon fibers—when, of course the fibers are not exceptionally strong; their virtue lies in their stiffness. There is some danger of a backlash from this kind of overselling.

POLYMERIC SUPERFIBERS

The inorganic superfibers we have described were developed mostly in the last 20 to 30 years. On the other hand, work on organic polymeric fibers of both high and low stiffness has been going on for at least 50 or 60 years. The first synthetic fibers were made from various forms of regenerated natural cellulose. Under names like *rayon* and *artificial silk*, they have had considerable success in the textile industry for clothing, although their application to "high" technology, which requires high strengths, has been limited.

The first important strong synthetic fiber to be developed was nylon, which was invented in the late 1930s by DuPont. As one might expect from its molecular arrangement, the Young's modulus of Nylon 66 is very low

Molecular structure of Nylon 66.

$$CO—CH_2—CH_2—CH_2—NH$$
$$NH—CH_2—CH_2—CH_2—CO$$
$$CH_2—CO—CH_2—CH_2—CH_2—NH—CH_2$$
$$CH_2—NH—CH_2—CH_2—CH_2—CO—CH_2$$
$$CO—CH_2$$
$$NH$$

indeed. On the other hand, nylon can be strained 20 percent or more before failure.

Nylon has been in great demand for women's stockings and other clothing in which its low stiffness and high extensibility are an advantage. Although the low Young's modulus has ruled out nylon for rigid applications in engineering and as a reinforcing fiber in composite materials, its extendibility and ability to absorb strain energy have rendered it very valuable for applications like parachutes and glider tow ropes.

A high capacity for strain-energy absorption can be a mixed blessing. It is very useful, for instance, when there is a need to "snatch" a glider off the ground. However, not long after World War II, the author was able to

Fabric materials make up the roof of the Schlumberger Research Center near Cambridge, England. (For other views of this structure, see pages 60 and 180.)

acquire a wartime glider tow rope for use as a warp for the anchor of his 40-foot sailing yacht. Crossing the English Channel one day we were becalmed, north of Cherbourg, about 20 miles off the French coast. Because the tides are strong in those parts I decided to anchor. We were in rather deep water, but we had the light, strong nylon warp to make, I hoped, the recovery of the anchor fairly easy.

When the wind came back the next morning, we began to wind in our nylon anchor rope, only to find that the anchor was "foul" of some obstruction, probably a wartime wreck. According to the classical doctrines of seamanship we would have to "sail the anchor out." That is, we would have to tack the ship against the wind, which was now quite fresh, hoping to wrench the anchor from its obstruction.

We duly paid out the slack of the cable, got sail on the ship, and set off—quite fast. The nylon cable stretched and stretched, as though it were the rubber of a catapult. Like a catapult it suddenly recoiled when it came taut, and flipped the ship astern with such violence that I thought we would lose our mast. In fact the mast survived—just. So we had to cut the rope and abandon the anchor. We sailed away for France, sadder and wiser. Let others be warned.

The *aramid* fibers are exceptional in having higher Young's moduli than other synthetic organic fibers. The best-known of these fibers is Kevlar, which has been developed by the DuPont Corporation and is used for tires, belting, and sails, among other things. Although the specific modulus of Kevlar 49 is a good deal lower than that of carbon or boron fibers, it is still respectable, and the specific tensile strength is higher for Kevlar fiber. This fiber is a good deal tougher than most of the other superfibers in tension, but it is sometimes apt to split or delaminate in compression.

Molecular structure of Kevlar.

THE NEW COMPOSITE MATERIALS

The fibers and superfibers that we have been discussing are, of course, designed principally for use as the strong and stiff components of advanced composite materials. Within the composite the fibers are bonded together by some sort of adhesive or matrix. Sometimes this is a metal, but more often it is an organic resin. Resins are lighter than most metals and they adhere better to the fibers. They are also more easily shaped: modern resins such as polyesters and epoxides can be molded and hardened at low pressures and moderate temperatures.

Although these resins are necessary to glue the fibers together, they do not contribute much to the overall stiffness of the material, and they do add quite significantly to the weight. It is therefore desirable to keep the proportion of resin matrix to a minimum. In practice the matrix generally accounts for between 30 and 40 percent of the total weight of a composite material.

It is scarcely practicable to tabulate elaborate sets of "typical mechanical properties" for the new composites. In theory the whole point of such materials is that unlike metals, they do not *have* "typical properties," because the material is designed to suit not only each individual structure, but each place in that structure. At least, that is the goal.

In traditional engineering, a single material, such as mild steel, is often used throughout the whole structure. Metals like mild steel are not only consistent from one piece to the next, but they are isotropic: that is, their properties are the same, or almost the same, in all three dimensions. The use of a single homogeneous material may make things easier for designers and managers, but it is not the way Nature works.

Nature, which utilizes fibrous materials which are in many ways analogous to the new composites, seldom installs isotropic materials in a living structure. In animals, the bones and tissues which carry the loads vary considerably from place to place, according to their function.

The problem of determining the directional properties of fibrous composites has attracted a lot of attention from mathematicians over a number of years. Their work bears out the experimental (and common-sense) conclusion that it is scarcely practicable to make a fibrous composite material which is truly isotropic—in other words, one that has identical properties in all three directions. When this result has been approximated in the laboratory, the mechanical properties have been undistinguished.

For some purposes, the best results are obtained when all the fibers are parallel, much like wood. Although such unidirectional materials, like wood, tend to be unduly weak across the grain and to split easily, there are actually a fair number of applications for parallel-fiber composites. A composite that is isotropic in two dimensions—that is, in the plane of the sheet—will have one-third the strength and stiffness of an all-parallel arrangement. For more exacting applications such as aircraft wings, where the best combination of spanwise and torsional stiffness is called for, such an arrangement of the fibers is not considered good enough, and more efficient and sophisticated orientations are constructed. But there is really no single, widely applicable, best design for a composite material. As we have said, the solution is likely to vary not only between different structures, but from place to place in the same structure as it does in animals.

The best design will depend on the nature of the loads the structure has to withstand, on the structure loading coefficients (Chapter 3), and so on. A lightly loaded structure calls for a different material than does a heavily loaded one. The scale of the structure is also important: a large structure will need a higher work of fracture than a smaller one. Sophisticated work-of-fracture mechanisms can be designed into composite materials, more or less to suit the requirements. Moreover, as we have seen in the case of wood (in Chapter 7), very effective toughening devices can be combined with cellularization to optimize the strength of a structure in bending and compression.

MATERIALS AND STRUCTURES OF THE FUTURE

It would take a bolder writer than I to venture to predict where all these new materials and new ideas are likely to lead us. The future will certainly look different, because engineers are being forced to rethink many of their fundamental presuppositions about materials and structures. The old cozy metallic world of the traditional engineers may yet last for a while longer, but in many industries it is increasingly seen to be open to question. It is very likely that within the next generation quite a number of major technological citadels will fall, and the social and economic consequences will be considerable. For one thing, both in design and in manufacture, the new products will need considerably higher levels of intelligence and skill than prevailed in the world of Henry Ford.

Many of the advanced materials now being developed are intended for a special market, defense and aerospace, where the requirements are often different from those of ordinary civilian goods. In aerospace, high stiffness—especially high torsional stiffness—is at a premium. This need is being met by composite materials based on the rather exotic superfibers, which tend to be expensive.

As we have mentioned several times, Nature nearly always manages to avoid an excessive requirement for torsional stiffness. Usually torsional deflections are resisted by means of *active elasticity*, that is, by muscular contractions triggered by nervous impulses. Where stiffness is needed to resist bending and compressive loads, as in trees and other plants, Nature

Riyadh International Stadium exemplifies new materials and new methods of construction. The teflon-coated fiberglass membrane roof of the stadium (1985) is 945 feet across and covers 60,000 spectators. A 440-foot-wide hole in the center exposes the open playing field, as required by international regulations.

solves the problem by reducing the density of the material by incorporating holes of one kind or another. Holes are cheap, both in manufactured and in naturally occurring structures: that is, they are efficient both financially and metabolically. In the development of the new synthetic materials, more attention might be given to the provision of "sophisticated" holes.

Only recently have materials scientists begun to take the idea of active elasticity seriously. Piezoelectric materials—materials that show strain in response to an electrical impulse—have been known to electronic engineers for a long time, but their structural implications are only beginning to be considered. If piezoelectric structural materials can be developed—and also reliable means of controlling them—then yet another structural revolution will be upon us. Yet, like many revolutions, it will be a reversion to the past: after all that is the way in which animals work.

Notes

Chapter 1

PAGE 1   EPIGRAPH
Aristotle *De partibus animalium* 1.5 (645a). Translated by A. L. Peck, Loeb Classical Library (Cambridge, Mass., Harvard University Press, and London, William Heinemann, 1937).

PAGE 3
For an account of academic attitudes, see Sir Eric Ashby, *Technology and the Academics: An Essay on Universities and the Scientific Revolution* New York, St. Martin's Press, and London, Macmillan, 1958).

PAGE 4   DEFINITION OF STRUCTURE
See, for instance, A. J. S. Pippard and Sir John Baker, *The Analysis of Engineering Structures* (4th ed., New York, American Elsevier, 1968).

Chapter 2

PAGE 11   THE PARTHENON
When traveling in Greece a few years ago, I met an American engineer who told me that he had been spending a good deal of time making measurements of the Parthenon, using Invar tapes and other modern

devices. He told me that the accuracy and symmetry of the structure were such that he did not see how they could have been achieved by the ancient Greeks, for the errors were less than the variations that would have occurred in the lengths of ordinary wooden or metal measuring rods due to thermal expansion.

PAGE 12   ARCHIMEDES
Plutarch *Marcellus* chapter 17.

PAGE 12   HEIGHT OF *INSULAE*
Strabo *Geography* 5.3.7 (C235). I take this account from J. Carcopino, *Daily Life in Ancient Rome* (London, Routledge, 1941), page 25.

PAGE 14   MEDIEVAL CATHEDRALS
Professor Jacques Heyman of the Department of Engineering at the University of Cambridge has analyzed the structures of the medieval cathedrals very extensively. See, for instance, his article, "Beauvais Cathedral," in *Transactions of the Newcomen Society* (London), volume XL, 1967–68.

PAGE 15   CARDANO AND LEONARDO
S. P. Timoshenko, *History of the Strength of Materials* (New York, McGraw-Hill, 1953), chapter 1.

PAGE 19   X-RAY DIFFRACTION
J. T. Norton and B. R. Loring, *Welding Journal*, Research Supplement, June 1941.

PAGE 21   GEORGE STEPHENSON
The writer's great-great-grandfather, Thomas Shaw Brandreth, F.R.S., (1788–1873), a director of the Liverpool and Manchester Railway, was responsible for hiring and supervising Stephenson for the railway company. Stephenson was a difficult personality, innumerate, illiterate, quarrelsome, and inclined to drink.

PAGES 21–22   FRANCE IN THE EIGHTEENTH CENTURY
For a contemporary French opinion, see Voltaire's *Lettres philosophiques*, (1734), especially Letter no. 10, "On Commerce."

Chapter 3

PAGE 37   EPIGRAPH
Rudyard Kipling, "In the Neolithic Age."

PAGE 45   HEDGEHOG
See P. Owers and J. Vincent, "Mechanical Design of Hedgehog Spines and Porcupine Quills," *Journal of Zoology* (London), volume 210 (1968A), pages 55–75.

PAGES 53–54   GIANT SQUID AND OCTOPUS
There are a good many dubious stores about giant squids, but there is a reliable record of an octopus (*Octopus gigantus verrill*) that had a total tentacle span of more than 180 feet (54 meters). Found dead in Florida in 1897, it was measured and photographed and some of its tissues have been preserved. See Jacques-Yves Cousteau, *Octopus and Squid* (London, Cassell, 1973), page 217.

PAGE 56   CIRCULATORY THEORY OF LIFT
Lanchester's book is *Aerodynamics* (London, Constable, 1907).

PAGE 70   CONVERGENCE OF STRUCTURAL FORMS
D'Arcy Wentworth Thompson's classic *On Growth and Form* was reissued by Cambridge University Press in 1959.

Chapter 4

PAGE 73   EPIGRAPH
Ecclesiastes 9:11.

PAGE 74   GRIFFITH'S PAPER
The paper appeared in the *Philosophical Transactions of the Royal Society* (London), volume A221 (1920), page 163.

PAGE 78   TENSILE STRENGTHS OF GLASS RODS AND FIBERS
See the article by J. G. Morley in the *Proceedings of the Royal Society* (London), volume A282 (1964), page 43.

PAGE 78   WHISKER CRYSTALS
Herring and Galt published their findings in *Physical Review*, volume 85 (1952), page 1060.

PAGE 78   STRENGTH OF NONMETALLIC SUBSTANCES
My paper appeared in *Nature*, volume 179 (1957), page 1270.

PAGES 79–80   H.M.S. *WOLF*
For these experiments, see, for instance, volume 1, chapter 31 of *The Design and Construction of Ships* by Sir John Biles (London, Charles Griffith, 1923).

PAGE 81   KOLOSOFF
A German version of Kolosoff's article appeared in *Zeitschrift für Mathematische Physik* in 1914.

PAGE 81   INGLIS
The paper was published in *Transactions of the Institute of Naval Architects*, volume 55 (1913), page 219.

PAGE 82   CRACK PATTERNS ON THE SURFACES OF GLASS
See the article by J. E. Gordon, D. M. Marsh, and M. E. M. L. Parratt in *Proceedings of the Royal Society* (London), volume A249 (1958), page 65.

PAGE 85   CONTROLLED TENSION
Marsh described his work in *the Journal of the Science Institute* (London), volume 38 (1961), pages 229–234.

PAGE 86   STEPLIKE SURFACES
See Marsh's article in the *Philosophical Magazine* (London), volume 5, page 1197.

PAGE 97   TIP OF A CRACK
See J. Cook and J. E. Gordon, *Proceedings of the Royal Society* (London), volume A282 (1964), page 508.

Chapter 5

PAGE 103   EPIGRAPH
The epigraph is from Samuel Butler's "Hudibras," part I, canto III.

PAGE 104   OLYMPIAN GODS
See Gilbert Murray, *Five Stages of Greek Religion* (Oxford, Oxford University Press, 1930).

PAGE 107   TABLE
Anthony Kelly, *Strong Solids* (Oxford, Oxford University Press, 1966).

PAGE 108   TAYLOR AND SAILING
Sir Geoffrey Taylor was a distinguished yachtsman and, among other things, was the inventor of the C.Q.R. or Ploughshare anchor.

PAGE 109   FRANK-READ SOURCE
A. H. Cottrell, *The Mechanical Properties of Matter* (New York and London, John Wiley, 1964), page 279.

PAGE 111   BATTLE OF MARATHON
P. H. Blyth, "The Effectiveness of Greek Armour against Arrows in the Persian War (490–479 B.C.)," Ph.D. thesis (unpublished), Reading University.

PAGE 119   KIPLING AND FATIGUE
The story "Bread upon the Waters" appeared *The Day's Work* (London, Macmillan, 1898).

PAGE 135   SMITH AND FORD
Adam Smith, *The Wealth of Nations,* opening paragraph of Chapter 1.
Henry Ford, *My Life and Work* (first published in 1922), Chapter 7.

Chapter 6

PAGE 159 ELASTICITY OF BIOLOGICAL MATERIALS
J. Vincent, *Structural Biomechanics* (London, Macmillan, 1982).

PAGE 160 MECHANICAL PROPERTIES OF LIVING TISSUE
H. Yamada, *The Strength of Biological Materials* (Baltimore, Williams and Wilkins, 1970).

PAGE 163 STRESS-STRAIN CURVE OF RUBBER
L. R. G. Treloar, *Physics of Rubber Elasticity*, 3d ed. (Oxford, Oxford University Press, 1975).

PAGE 166 STRESS-STRAIN CURVE OF SOFT TISSUES
J. E. Gordon, "Mechanical Instabilities in Biological Membranes," *Proceedings of the Institute of Comparative Physiology, Ancona, 1974* (New York, American Elsevier Publishing Company, 1975).

PAGE 170 ELASTICITY OF BIOLOGICAL SOLIDS
See my article referred to just above and S. A. Wainwright, "The Role of Materials in Organisms," in the same volume.

PAGE 170 ELASTICITY OF ELASTIN
J. M. Goseline, "Hydrophobic Interaction and a Model for the Elasticity of Elastin," *Biopolymers*, volume 17 (1978), page 677–695.

PAGE 173 PAIN IN A TECHNOLOGICAL STRUCTURE
See, for instance, *Proceedings of the DARPA/AFML Review of Progress in Quantitative Non-Destructive Evaluation* (Thousand Oaks, Calif., Rockwell International Science Center, 1980).

PAGE 176 BALD VENUS
J. G. Landels, *Engineering in the Ancient World* (London, Chatto and Windus, 1978), especially Chapter 5.

PAGE 178 EFFICIENCY OF BICYCLE RIDING
G. Goldspink, "Biochemical Energetics of Fast and Slow Muscles," *Proceedings of the Institute of Comparative Physiology, Ancona, 1974* (New York, American Elsevier Publishing Company, 1975).

PAGE 179 GREEK WARSHIPS
J. S. Morrison and R. T. Williams, *Greek Oared Ships*, 900–322 B.C. (Cambridge, Cambridge University Press, 1968).

Chapter 7

PAGE 164 MORPHOLOGY OF CELLULOSE FIBRILS
See the article by D. R. Cowdrey and R. D. Preston in the *Proceedings of the Royal Society* (London), volume B166 (1966), page 245–272.

PAGE 170   WOOD CELL LOADED IN TENSION
See the article by G. Jeronimidis and me in *Nature* (London), volume 252 (1974), page 116.

Chapter 8

PAGE 181   EPIGRAPH
Rossetti's advice appeared in a letter to Hall Caine and is quoted in Caine's *Recollections of Rossetti* (New York, Gordon Press, n.d.).

Sources of Illustrations

PAGE 160
Jim Brandenburg

PAGE 161
Saint-Cloud, Eugène Atget. Collection, The Museum of Modern Art, New York. The Abbott-Levy Collection. Partial gift of Shirley C. Burden

PAGE 162
National Aviation Museum, Ottawa

PAGE 163
Collection, The Museum of Modern Art, New York. Philip Johnson Fund

PAGE 164
Institute of Paper Chemistry

PAGE 165
Philip Rosenberg

PAGE 173
John Shaw

PAGE 174
Trestle Work, Promontory Point, Andrew J. Russell. The Oakland Museum

PAGE 175 TOP
Susan Middleton

PAGE 175 BOTTOM
John Currey

PAGE 176
Andreas Feininger/Life Picture Service

PAGE 177
George Bernard/Oxford Scientific Films

PAGE 178
The Monarch of the Glen, Edward Landseer, courtesy of Guinness PLC

PAGE 180
Alastair Hunter

PAGE 183
The Gare Saint-Lazare, Claude Monet, The Art Institute of Chicago

PAGE 185
from *Plastics,* by Sylvia Katz, Thames and Hudson Ltd., London, 1984. Photograph by Sylvia Katz and John Kaine

PAGE 194
American Matrix Inc.

PAGE 197
Tsu-Wei Chou

PAGE 198
Alastair Hunter

PAGE 202
Horst Berger

Index